GARAGEBAND BASICS

THE COMPLETE GUIDE TO GARAGEBAND

AVENTURAS DE VIAJE

Illustrated by
NEIL GERMIO

Copyright SF Nonfiction Books © 2021

www.SFNonfictionBooks.com

All Rights Reserved
No part of this document may be reproduced without written consent from the author.

WARNINGS AND DISCLAIMERS

The information in this publication is made public for reference only.

Neither the author, publisher, nor anyone else involved in the production of this publication is responsible for how the reader uses the information or the result of his/her actions.

CONTENTS

PART I

Introduction	3
Glossary	5
GarageBand Basics	17
Projects	25
Recording	31
Media	39
Arrangements	49
Controls	64
Enhancements	79
New Skills	83

PART II

Introduction	91
Part 1 Recap	93
Projects and Recording	103
Arrangements	108
Controls	116
Mobile GarageBand	129
Tips and Tricks for Mac GarageBand	165
Conclusion	175
Shortcuts	178
References	184
Author Recommendations	187
About Aventuras	189

THANKS FOR YOUR PURCHASE

Did you know you can get FREE chapters of any SF Nonfiction Book you want?

https://offers.SFNonfictionBooks.com/Free-Chapters

You will also be among the first to know of FREE review copies, discount offers, bonus content, and more.

Go to:

https://offers.SFNonfictionBooks.com/Free-Chapters

Thanks again for your support.

PART I

INTRODUCTION

So you've decided that you want to start recording and producing your own sound. You've done some reading and your research has led you here. This GarageBand guide will help you start your journey and help you achieve your ultimate goal of wonderful sound production. Whether you are using instruments or recording your own lyrics or podcasts, this guide will help you achieve it all.

The first part of The Complete Guide to GarageBand will cover all the basic knowledge you will need to be able to start the use of GarageBand. The second part will cover in-depth tutorials to create a project from start to finish. Part 1 is beginner-friendly and is easy to use no matter what your skill level is. Whether you have used GarageBand before or it is the first time that you're using it, this book will help guide you to production perfection.

In this guide, we will cover the basic usage of GarageBand. Each chapter will offer mini-tutorials on the user interface and how to use the software program. You will also be able to follow the mini-tutorials to get used to navigating the controls and shortcuts offered within the program. Each chapter will also include any related content that can be found in other chapters within this guide.

While GarageBand can be used solely for music production, you can also use the software to add music to a short film or record your own podcast. The software has limitless possibilities and will allow you to produce magnificent sound no matter what your skills are. All chapter tutorials can be used to grow your knowledge and experience when it comes to using the software and creating amazing sound.

The guide covers GarageBand basics that include navigation within the program as well as shortcuts, projects, and how tracks link into them. Recording, whether that includes recording voice or instruments or using the in-program sounds. Media, and how linking sound to your media projects can be achieved. Arrangements and how to best arrange regions, as well as editors and music notation. Controls and how they influence your sound. Enhancements and how tempo and transposition can better the sound you create and lastly this guide will cover one of the most useful parts of Garage-Band, the Lessons section.

Please note that there is a glossary added to this guide to help you understand terms that you have not come across before. It will be at the end of the introduction for ease of use. After the conclusion at the end of the book is a shortcut guide. The shortcut guide will offer you all the possible shortcuts you may need when using GarageBand.

You are finally ready to begin the amazing journey to becoming a fantastic sound producer, and it all starts here!

GLOSSARY

Amp (Amplifier): Usually used along with electric guitars. You can use either a stack or combo amplifier. GarageBand offers a variety of amps to create a fuller sound.

Apple loops: Apple loops are pre-recorded sections of audio that can be used in your projects. These loops are created for looping and repeating. GarageBand has a variety of loops that you can use wherever you wish to in your project. You can also adjust the tempo and pitch of any of these loops.

Arrangement region: Rectangle segments of 8-bars. These regions can be added to create various different sections in your project, including but not limited to verse, intro and chorus. Arrangement regions can be used to quickly try out new arrangements when you've already added material like other recordings or loops. They can be moved around in your track for ease of use.

Audio: The sound that you will be recording or transferring. It is the sound you will be recording via your microphone or the software instrument section. This will appear as the audio track section in the right-hand section.

Audio interface: This device converts analog signals (the sound your microphone records or the sound your instrument makes) into digital data that can be processed by GarageBand and your computer. It also works the other way around, converting digital audio data into analog signals before pushing it through your speakers.

Audio Units (AUs): This unit is considered the standard for any plug-ins into the OS X format. It can be used in conjunction with any of the audio effects, generators, and software instruments. This formatting is built into GarageBand and any extra software can access this installed AU plug-in. It is important to note that GarageBand offers support for all AU format plug-ins.

Automation: This feature allows you to make changes over the duration of your project. Automation curves for individual tracks and master tracks are included in the GarageBand interface. The automation curve adjusts the volume, tempo, and various other settings including the pan of a track when you add these as automation points onto the automation curve. These settings can be changed by dragging these points.

Bar: A group of beats that are heard together. The time signature in the project will show how many beats each bar contains, it will also show you the musical value of each beat. When using the music notation section, bars will be separated by vertical lines.

Beat: The rhythm used when basing the pulse of a piece of music. This will be a regular pulse. The time signature in the LCD will also show you the number of beats per bar and the note value that each beat will have.

bpm (Beats per minute): This indication provides the tempo of any piece of music.

Chord: Notes that are played together are called a chord. They are considered minor or major and can have a seventh note added. If you are using a software instrument track the note you play will be shown on the LCD screen at the top of GarageBand.

Chorus: Can be defined in 2 different ways. (1) a part of a song that is generally repeated between various verses. (2) an effect that is made by having multiple voices or instruments repeat a sound at slightly different timings.

Clav (Clavichord): A keyboard instrument. Similar to a guitar but instead of striking the chords the user plucks at the strings. It is much closer to a guitar than a piano. GarageBand offers various different clav sounds and track selections.

Clip, clipping: Clipping is found when the track volume level is higher than the volume that your speakers or equipment can handle and produce. GarageBand's track header has a volume meter that will become red if clipping does occur on any of your tracks.

Compressor: A compressor evens out the sound between the really soft and extremely loud sounds. This helps create a smoother overall sound. Compression creates a sound that doesn't lose its quality if it is played over equipment with a narrow sound range. GarageBand offers a built-in compressor on all the tracks including the master track.

Control bar: The control bar is found in the tracks section of the interface. Here you can change and alter what you see on your screen while working in GarageBand.

Core audio: The core audio is a standard audio driver that is used within all Mac computers that are currently running Mac OS X. This driver system allows access to all the audio interfaces that are a part of the core audio used by GarageBand. Any hardware that is Core Audio compliant is compatible with GarageBand.

Core MIDI: The core MIDI is a standard driver that is used within all Mac computers that are currently running Mac OS X. This driver is extremely important as it allows you to connect any MIDI device, without this driver software you will not be able to connect MIDI to your computer and use it within GarageBand. GarageBand is also compatible with any MIDI hardware that has the Core MIDI component.

Count-in: The metronome beat sounds that are made before the recording of an instrument or vocal track starts. This is typically used for one bar. The count-in helps you stay on tempo with the project.

Cycle region: The cycle region is used to record over a specific part of a track. To be able to use the cycle region functionality you will have to toggle it on for the selected track. The player will return to the start of your selection when it reaches the end of your selection.

Decibel (dB): This is the standard method of measuring the volume or the loudness of a specific sound. 1dB is the smallest volume change the human ear can hear.

Distortion: Distortion can be split into two definitions. (1) A sharp sound that is found when the volume exceeds the volume that equipment can produce. (2) an effect that can be used to create a rough and loud sound.

Drummer: A virtual drummer sound that can be added by using the Drummer track. It can be altered in the Drummer Editor to create the perfect sound for your project.

Drummer editor: The editor in GarageBand is used to edit the playing style as well as the sound that is used for the virtual drummer track you have added to your project. You can change the settings surrounding the kit and the fill settings.

Dynamic: Dynamic can be defined in two different ways. (1) refers to the difference found between the lowest and higher volume of a specific track. (2) A change that occurs over a period of time.

Echo: An effect that can be found when a specific sound is played over and over. It is generally only noticed when a track is in the playback stage. This sound makes it seem as if the sound is being played in a large space rather than a small studio. It is also often referred to as the delay.

Editor: The area for either the drummer, software instrument, or vocal track to be edited or changed. You can edit regions or loops in the editors. The editors are found below the track area. Edit changes are different depending on the track selection.

Effect: Software or hardware that changes or alters the sound of any given track. GarageBand offers various effects that can be added to tracks like; compressor, echo as well as EQ and reverb.

EQ (Equalizer or Equalization): It is an effect used to change the sound frequencies of a selected track. The EQ can be used to create a range of dramatic or subtle changes to the quality of your track.

Fade-out: Fade-outs are used to gradually soften or lower the volume of a selected track. This is usually done at the end of a song. GarageBand has the ability to add this automatically to the track in the tracks curve.

Filter: Filter is an effect that only allows certain frequencies to be heard, while others are blocked out. GarageBand offers an Autofilter that automatically does this, but you can opt to use various EQ effects to create the same effect.

Flanger: The flanger is an effect that deepens the sound of your track by replaying the same sound repeatedly. These repeats are slightly out of tune when compared to the original. GarageBand offers a flanger effect as well as other flanger effects when using any Electric Guitar tracks.

Flex time: Flex time allows the user to edit any timing for both the notes and the beats in a track. GarageBand allows you to add flex markers into your tracks inside the editors, on the waveforms. By clicking the peak the editor will add a small flex marker to the waveform where the flex time can be added. Flex markers are usually used for compression, stretching, and expanding audio sections.

Fuzz: Fuzz is generally a distortion effect added to Electric Guitars to create a thick and rough sound. GarageBand has various effects that can be added to create this sound.

Gain: Gain, similar to volume in a sense, lowers or raises the level of electronic audio signals. This is done by adjusting the gain on an amp. By changing the gain on the amp you will either create a clear and light sound, or a loud and distorted one.

Grid: The grid in GarageBand can be used to align and adjust the tracks. It will also align any beats if you wish to use them like that. Turning on the grid will snap the following things to the nearest grid position: any loops, moving and resizing of regions, the playhead moving, as well as automation point movement and the cycle regions adjustment.

Groove track: The groove track is used to match or sync the timing from the other tracks. When using a groove track, all other tracks will adjust their time to sync to the groove track. It is also important to note that only one track can be allocated as the groove track.

Input source: This setting is used when you are recording sound from a microphone or an instrument. It will be the selection you make to enable you to hear the sound.

Intro (Introduction): Usually the beginning part of a song, and it will generally last until the first verse.

Key: The key is the terminology used for scales, or the series of notes that generally start at a specific pitch and go higher, or lower. The starting note is usually referred to as the root note. GarageBand offers key selection at the start of the project, but you can also change it in the LCD during the project.

LCD (Liquid Crystal Display): The LCD is the small box in the top middle of the screen that shows the scale, tempo, and the key signature for the project. You can change the outputs that the LCD shows depending on the selections you make.

Library: This section is located in the left pane of GarageBand. This will be the section where you choose the different tracks you want to add to your project.

Lock (Tracks): This feature allows you to lock tracks. Locking a track means that no changes, editing, recording, or new regions can be made with or for that track. You will have to unlock the track to be able to do any of these actions. The lock functionality is found in the track header.

Loop: Loops can be described as (1) a MIDI region or pre-recorded sound that was created to be repeated in a loop or (2) copying or duplication of a selected region so it plays multiple times.

Loop browser: The area GarageBand offers where you can search for loops either by using the search function, by filtering, or by adding your own loops. You can customize the way you view the loops in the loop browser.

Master track: The master track is the track that controls the entire sound for the track before it plays through speakers or headphones. Any effects or changes made to the master track will affect any of the other tracks and their output. You can edit the master track in the editor window with the smart controls.

Metronome: The metronome functionality in GarageBand allows you to measure the time intervals between beats. The metronome makes a clicking sound to indicate each interval. GarageBand uses the metronome during any form of track playing or recording. The metronome functionality also offers the user a Count-in for recording purposes.

MIDI (Musical Instrument Digital Interface): This is a standard protocol that producers and musicians use to connect their electronic instruments to their computers. The MIDI uses an interface to translate the music data to the computer. They are generally USB compatible.

Mix, mixing: Mixing is considered the action where you mix or bled sections of tracks together to create a more cohesive track. Mixing is done by changing the volume on each track, adding panning to them, and even adding effects. You can also do this by making adjustments to the master track, or use the automation functionality to generate or change the mix of various tracks.

Monitor, monitoring: Monitoring is done when you are hearing the sound a specific instrument or vocals make while you are busy recording in GarageBand. Turning the monitoring off when you are using a computer's speakers instead of headphones will allow you to avoid any feedback.

Mono, monophonic: Monophonic devices are generally microphones or instruments that only have one output. Devices with outputs that are located on the left and right are called stereophonic.

Make sure that you select the correct source when adding in an instrument or microphone.

Movie track: The movie track is the track that displays both frames from the movie and the audio track for that specific movie. The audio will play and the frames will change.

Notation: The notation is considered a visual view of what music sounds like. It includes the notes, the rests, and any other music symbols. GarageBand offers an editor where you can view the software instruments as music notation.

Note: Notes are the music terminology for the pitch of a sound. Common scales like Minor and Major all have 7 basic notes, ranging from A to G. Chromatic Scales however have more than the Minor and Major notes, as they have 12 notes.

Octave: Octaves are measured as the pitch difference between two notes, they are either twice the pitch of one note or one-half the pitch of other notes. Notes that sound similar are considered to be an octave apart and they are generally considered to be the same note when on a scale. You find 12 semitones between notes that are an octave apart.

Outro: The ending of a song, generally fading out into silence.

Overdrive: An effect that can be used to create a sound similar to a tube amplifier. When raising the gain on this effect it will create a distorted sound on your track. GarageBand offers an overdrive effect as well as an overdrive stompbox effect, specifically for Electric Guitars.

Pan (Panorama): The pan is considered the space between 2 speakers in a stereo field. The track's sense of direction, and where the sound is coming from is determined by the pan of a track. The knob in the track header allows you to adjust the pan of that specific track.

Patch: Patches are complete sections that can be added to track that include the sound, any effects, and the settings concerning routing. Patches can be found in the library or through third-party downloads.

Phaser: Phasers are known to create a whoosh sound, similar to a plane or jet passing by your house. It is created by playing the same section repeatedly but having the phase change between versions.

Pitch: The lowness or highness of a specific note or sound. Pitch also corresponds to the frequency of a sound.

Playhead: A small arrow icon attached to a line that runs from the top of the tracks area, all the way to the last track. It can be used to stop and start the track at a specific time. The arrow icon indicates where the playhead is currently located.

Project Chooser: The selection area when creating a project for the first time. Here you can create a new project or open a project that has been used recently. There are different templates to choose from, and you can open the Lesson section from Apple's Learn To Play functionality.

Region: A rectangle section of music that can be considered a loop, any sort of recording, sound, or media. They can be used multiple times or only once. Regions are different colors depending on the track selection you made and the region selection you made when importing or adding the region. Regions can also be edited and changed in the editors' area.

Remix: Remixes are created by taking an old song and adding effects, loops, regions, and other media to change the sound and feel of the track.

Reverb (Reverberation): Reverb is used to create different feels for sounds. This means that the track could sound like it is being played within a large concert hall, as opposed to a studio or a bar. Each track has a master reverb effect that can be altered to the desired effect.

Ruler: A ruler is a tool that runs across the screen above the tracks area. It allows you to align, merge and connect regions or segments much easier but also indicates the time in the project.

Scale: The term used to portray a selection of notes that form a melody, a chord, or an entire soundtrack. There are Minor and Major scales, which are the most commonly used. The scale can be set at the beginning of a project or in the LCD during the project.

Screen control: The screen control is used to change aspects of the track and its sounds. GarageBand's screen controls have labels to help you better understand the changes you are making to the selected track.

Semitone: Semitones are the smallest sections of difference between two notes. 12 semitones are found between two octaves.

Smart control: Smart controls are a selection of controls created by GarageBand to allow you to better control the changes you make to tracks. They can make changes to tracks or plug-ins.

Software instrument: Recording a sound using a MIDI keyboard is grouped into the software instrument track selection. The MIDI plugin translates the musical data into sound data for GarageBand to use. GarageBand also offers software instruments for use in your project if you do not have an instrument or a MIDI plug-in. When using either an actual instrument or a software instrument, the section will be labeled as regions.

Stereo: Instruments or microphones with left and right outputs are considered to be stereophonic. Single output devices are called monophonic. As with monophonic, make sure of the correct selection when plugging in your instrument or microphone.

Stompbox: Stompboxes are pedals that are used by guitarists to create special effects and sounds. GarageBand offers various stompbox selections that can be added or edited for your project.

TAB (Tablature): The musical notation that shows a player where they should be putting their finger to create a specific note or sound.

GarageBand's Lessons sections allow you to see notes and chords in TAB or music notation form.

Take: As with filming, take is linked to a series of recordings that are done in one grouping. Musicians and producers oftentimes do multiple takes of a song, or sound and then choose the one that created the best sound. The cycle region in GarageBand allows you to record multiple takes and then choose the recording you liked the best.

Tempo: The speed of the beats in the track or song you are creating. It is normally measured in bpm (beats per minute). You can select this at the beginning of the project or change it during the process in the LCD.

Time signature: The time signature indicates how the time of your tracks is split between bars and beats. GarageBand offers you the capacity to select this at the beginning or change it during the project in the LCD. It is split between two numbers. The first number indicates the beats that are found in each bar and the second shows the value of each beat.

Timing: The timing of a track is categorized by the events that are aligning with the bars and the beats of your track.

Track: Tracks are the sections next to the library where the regions are placed into. The block will have waveform patterns to indicate the sound as the volume of the track that you have added. Each track, depending on the selection you make, will be a different color. Audio tracks are blue, software instrument tracks are green and drummer tracks are blue. Any loops, recordings, or plug-ins will be the color of the track selection you made.

Track header: The track header is the area found between the tracks and the library. Here you can alter the track name, the icon, set the volume, pan, and solo the track.

Tracks area: The tracks area is the main screen you will be working on during your project in GarageBand. It is the entire window, including the library on the left and the editor below the tracks section. The track area includes all the areas you will need to work on your project and is also a great visual way of seeing your project come to life.

Transpose, transposition: Transposition is defined as the change in pitch to a specific region, project or track, by playing it in a different key. The pitch can be changed with the Master Pitch automation curve, or with the pitch slider in any of the editors.

Tune, tuner: Tuning is generally used with guitars when the pegs in the headstock are turned to adjust the pitch of a sound. GarageBand includes a built-in tuner that can be used to change the pitch of your instrument.

Velocity: Velocity is the pressure exerted on the keys when you are playing a keyboard, whether it is via USB or with a MIDI. Playing software instruments at a higher velocity will make the notes sound different than at a lower velocity.

Volume: Volume is measured in dB (decibels) and measures the loudness or softness of a sound. GarageBand allows you the functionality to change the volume of each individual track as well as the overall volume in the Master Volume, in the Master Track.

GARAGEBAND BASICS

Now that we have the introduction out of the way we can move on to all the basics you need to know. This section of the guide will cover all the supplies you will need as well as any tips and tricks on how to use the software.

While you can use extra instruments within your projects, GarageBand offers a variety of sounds that are pre-made that you can just plug into your project and produce amazing media. More on plug-ins and what they can do for you in chapter 2.

The first time you use GarageBand it will prompt you to download the essential loops and instruments. This will give you access to a large library of sounds that you can use in your projects. You can opt to not download them, but it will leave you with a significantly smaller selection. If you choose to record all your sounds there would be no use for these extra sounds. Unfortunately, if you choose not to download these sounds you will be prompted to do so every time you start GarageBand.

GarageBand will also offer you another download dialogue box that suggests downloading the entire sound library that GarageBand has to offer. The entire library is extremely large and will offer you almost 20 gigabytes worth of sound. If you wish to use the entire library I would suggest downloading it all.

First View

When opening GarageBand for the first time you will be greeted by the main screen titled "Choose a Project." On the left-hand side of the screen, there will be 5 menu items from which you will be able to choose.

The first menu item will be New Project. As the title suggests this menu option can and will be used when you are creating a new project. The second menu option that will be available for you to use will be Recent. Again, the name is rather self-explanatory. This menu will allow you to open other projects you have worked on. Please note that if you have not used GarageBand before you will not have any recent files. The third menu is called Learn to Play. This menu option, along with the next menu option, Lesson Store, will be covered in full in Chapter 8.

Apple offers GarageBand users the option to use their lesson store to learn how to play instruments like the piano, guitar, or keyboard. It is a wonderful addition if you wish to learn to play any of the instruments while also learning the basics of music and sound production.

The final menu option will be Project Templates. This menu option allows you to pick any of the templates that have been created beforehand. Similar to the templates you would choose when you are creating a text or slideshow document. The Project Templates section allows you to pick from 6 different collections namely: Keyboard Collection, Amp Collection, Voice, Hip Hop, Electronic, and Songwriter. Selecting any one of these collections will create a project surrounding your selection.

The main window will have a small block that is highlighted called Empty Project. This option can also be selected as it creates an empty project, with guided prompts from GarageBand on how to use this feature. Underneath the Empty Project box, you will see a small section that has the title Details. This section covers the Tempo, Key Signature, Time Signature as well as the Input and Output Devices. For now, we will leave these settings as the defaults.

For this chapter, we will start with the Empty Project option. By selecting this option you will give guided prompts on how to continue. After double-clicking on the Empty Project section the GarageBand automated guide will pop up. It will prompt you surrounding the type of track you wish to use. There are 3 different options for the track selection. These options are Software Instrument, Audio, and Drummer. You can select whichever is sufficient for you.

Each track allows the user different sounds and options within the GarageBand software. Software Instrument allows you to plug in a MIDI keyboard to allow you to play any of the instruments that are pre-programmed into the software. If you do not wish to plug in your instrument this is the option to select. You do not have to plug in a MIDI keyboard as you can set up your keyboard to do the same thing. Audio is split into 2 different selections, recording using a microphone or connecting a guitar or bass. If you are not connecting a stringed instrument you can select the microphone selection. The final of the three options is called Drummer. This selection allows you to add a pre-programmed drummer track that automatically syncs to the media you are creating and plays for you. This is especially helpful since you do not need to record the sound yourself and struggle to find the correct music notation and tempo.

We will be selecting the audio options that allow us to record using a microphone. This is the easiest option if you are looking for basic sound production. Once your selection has been made you will be taken to the Main Window as seen below.

Before we get into the basics that the software has to offer, the guide will be looking at the various buttons that can be found within the user interface. These buttons will help you navigate the screen a lot easier.

In the top left-hand corner, you will find an icon that looks like a filing cabinet. It hosts the various libraries of sounds and plug-ins that you can access within the software.

Next to it, you can find the question mark icon. This is the Quick Help section, similar to the Frequently Asked sections on websites.

The third icon that looks like a dial is the Smart Controls button. The smart controls can be accessed and changed with this specific button.

The scissors icon is used as the Editors button and will open the editing section for the selected tracks or project.

In the right-hand corner, you will find a notepad icon. It opens a notepad section where you can make notes about selected tracks or the project as a whole.

The final icon is located in the far right-hand corner and hosts the Looping functions for tracks and regions.

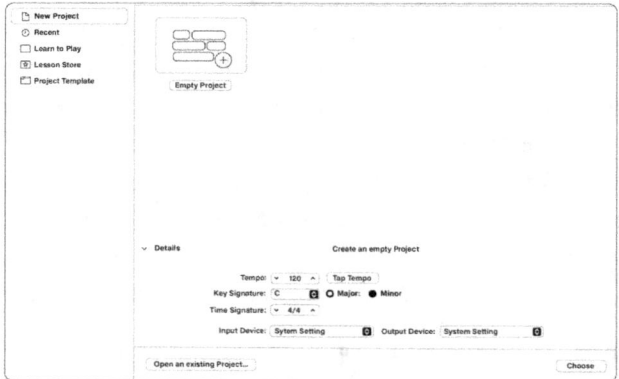

Basics

Now we can view the track screen and we can move on to using the basic interface and how to use it. As with any other software most of the actions within GarageBand do have shortcut keys that can be used. A chart will be added to the end of this chapter for convenient viewing purposes.

There are various basic keys that you will need when first starting out a project. Knowing how to pause and play as well as record will be a lot easier when you get around to using the shortcuts instead of having to use your mouse.

To save your project you will have to click on File and save. The save dialogue box will only appear the first time you save a file. Here you will be able to change the name of the file as well as select where the file needs to be saved.

To delete your project you can locate your GarageBand file in the Finder menu. You can simply drag it to the Trash. If you wish to recover it, you can do so if you haven't cleared your trash yet. If you have cleared your Trash you can unfortunately not recover your file so be careful when deleting and clearing your Trash out.

To open a recent file you can click on the File and then select Open. This will open the recent GarageBand dialogue and you will be able to select your file from the recent group. By double-clicking on the file you will be able to open it.

Now that you understand the basics surrounding the different keys you can and will be using we can move on to adding small sections of audio and recording them together. On the left-hand side, you will find the Library panel. This panel hosts all the different sounds and audio that come preprogrammed into GarageBand that you can use in your projects.

To use any one of the sounds in the library section, simply select it. It will automatically change the audio for you. To add more sounds simply press the plus icon above the track. This will bring up the track-type dialogue box where you can select which type of track you would like to add. Select whichever is appropriate and presto! This will leave you with two tracks that you can adjust and change as you wish. To delete a track, simply right-click it and select delete track from the menu.

Exporting your file is just as easy as it is to save it. Simply go to the share section and select whichever selection applies to you. You will be able to share your song to Apple Music, Soundcloud as well as share it over AirDrop and you will be able to send it over an email. You can even export your song to a disk. If you are not done editing or changing your file and you want to edit it on either your iPad or iPhone you can select the Project to GarageBand for iOS. This will create a .band file that you can airdrop to your other device and continue the editing process.

Shortcut Keys

Shortcuts will help you during your production phases. It will make the entire process much faster and will help you get basics under your belt rather quickly. While there are a lot of shortcuts I advise learning the basics like save, open, delete, record and playback.

It is important to note that there are different shortcuts for different areas of the software. There are 3 different categories of shortcuts. Main window shortcuts, editor shortcuts, and global track shortcuts. All 3 categories of the shortcuts can be found in this section of the chapter.

It is important to note that some of the shortcuts use numeric keypad shortcuts so it would be wise to invest in a full keyboard that has there or an extra numeric keypad plug-in.

Main Window Shortcut Keys

These shortcuts are used in the main window as referenced above.

In this section of the shortcut keys, only the most important shortcuts will be referenced. A full list of all the shortcuts for GarageBand will be added at the end of the guide for your convenience.

Shortcut	Action
R	Starts the recording
Space bar	Starts or stops the playback
K	Turns the metronome on or off
Command + Q	Quits GarageBand
Command + N	Opens a new project
Command + O	Opens an already existing project
Command + S	Saves the current project

Shortcut	Action
Shift + Command + S	Saves the current project as ...
Command + Z	Undo
Shift + Command + Z	Redo
Command + X	Cut
Command + C	Copy
Command + V	Paste
Command + A	Selects all

Editor Shortcut Keys

The editor shortcut keys are used when you are using any of the editors that GarageBand offers.

Shortcut	Action
Options + Space	Preview the currently selected audio
Options + Up Arrow Key	Transposes the selected notes up by one semitone
Options + Down Arrow Key	Transposes the selected notes down by one semitone
Options + Shift + Up Arrow Key	Transposes the selected notes up by one octave
Options + Shift + Down Arrow Key	Transposes the selected notes down by one octave

Global Track Shortcut Keys

These shortcuts are used to either hide or show any of the arrangement, video, or transposition tracks as well as the tempo tracks.

Shortcut	Action
Shift + Command + A	Shows or hides the arrangement tracks
Shift + Command + O	Shows or hides the movie or video tracks
Shift + Command + X	Shows or hides the transposition tracks
Shift + Command + T	Shows or hides the tempo tracks

Mini Tutorial 1

The following tutorial will help you learn the basics that have been discussed in chapter 1:

1. Open GarageBand and create an empty project.
2. Select the audio selection when prompted to do so.
3. Add any voice you prefer, listen to all of them before making your decision.
4. Use the shortcut key to save your file as My First Project 1.
5. Add a second tack that falls under the drummer track type.
6. Save your file using the shortcut.

This chapter relates very briefly to the content that can be found in chapter 5. Please check it out if you're looking for more information on song arrangements, music notation, and editors.

PROJECTS

When starting a new project it is important to note that different parts of each track need to be considered. Whether or not you're adding sound to media or creating an entire track for production purposes. It is also important that the properties of every project can be adjusted to help with the key, time signatures as well as tempo.

The project properties can be found on the main window when opening a new project. You can also change these properties at a later time when the sound you are creating changes. Sometimes you will have to set some of the properties beforehand, something like tempo drastically affects the playback of certain media files, and it is best to set these before starting the project.

There you will be able to change Tempo. The tempo can range from 5 to 990 beats per minute. While you can set the tempo up by one each time you can use the Tap Tempo button to select the tapping tempo for the project. The tap tempo will randomize the tempo for you.

The Key signature can also be changed here. You can select any of the 15 keys in the dropdown window, as well as select if it is in Major or in Minor.

The time signature will generally always be 4/4 by default but the range from 1/4 to 99/4 depending on what you need for your specific project.

The input device will primarily be the instrument, microphone, or headset that you have plugged in. If you opt to not have a mic or instrument plugged in I suggest always plugging in a good quality headset. That way you won't have any feedback when you are recording anything or playing anything back.

Like the input device, the output device will primarily be the system settings or the built-in output device. This is the section that saves and outputs your media to the file type that you choose at the end when you are exporting the file.

All of these things can be changed in the LCD when you are in the main window. In the top center of the main window, you will find a block that has the bar, beat, tempo, key signature, and time signature. By clicking on these elements you will be able to adjust them as you need to.

Tracks, Patches, and More

Tracks, patches, and plugins. All of these words look at the same thing. They all affect the overall sound and have a very powerful role to play. With all of these elements, you will be able to create a sound that is beyond your wildest imaginations. This section will cover everything you need to know about tracks and how to use them.

Tracks

Tracks allow you to organize and control the entire sound you wish to create. Whether it is an audio, drummer, or software instrument track. Audio tracks are generally used along with drummer and software tracks, to create your final track before linking it to the master track. The master track is the final step before completing your project and allows you to change the entire sound if you see fit.

Using the screen above as a reference, we can look at the three different audio tracks. The first track is the audio track, it is a Deeper Vocal audio recording.

The second track, Flying Circus Piano, falls into the software instrument category and can either be created by plugging your keyboard into your computer using a MIDI connector or you can set up your

GarageBand to allow you to use your computer keyboard to take the place of your instrument. We will discuss this in Chapter 3.

The final track as seen in the image above is the Drummer track. These tracks are automatically generated and there are various unique types of sounds you can choose from in the Apple library.

It is important to note that all three classifications can be imported from different media sources and used instead of opting for the Apple Library samples that you can use and create. If you do opt for using the Apple library sounds, you can change and alter each of the tracks in the editors of each of these tracks. Editing each track will be covered in chapters 3 and 5.

While loops and regions fall into the track category they will be covered in depth in chapters 4 and 5, respectively.

All tracks can be recorded, soloed, and muted during playback. For example, if you wanted to record just the drummer track, you would solo the sound by clicking on the headset icon in the drummer track. This will mute all the other tracks and allow you to record only the drummer track. On the other hand, if you wanted to mute the other tracks, you could use the speaker icon right next to the headset icon on all the tracks you wish to mute. This will allow you to add other sounds later on, without having to solo the drummer track.

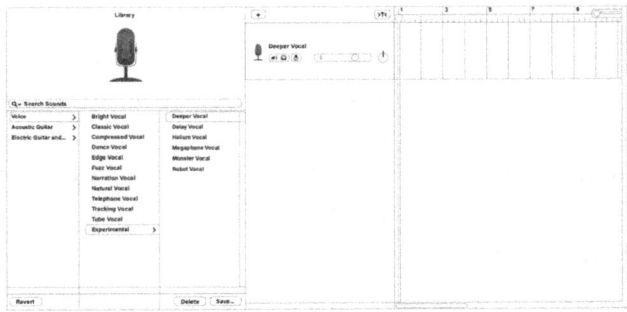

Patches

Patches are the specific sound and effect you want each track to make. These are found when selecting the track type and what type of effects you would like the track to have. The patches are found in the section attached to the library section on the left.

On the left-hand side, you will find the library. The first pane is considered the library. The second and third panes are considered the patches pane. This is where you would select the effect you want your track to have.

For example, when choosing a voice section for your audio track, the first section (the library) would have 'Voice,' the second section (the patch) would be called Experimental, and the third pane (the patch) would be called Deeper Vocal.

Plugins

Plugins can be used to further your sound and create a deeper sound. Any type of plugin can be found within the Apple library or you can find them from various other online sources. Plugins are generally used alongside any patches for specific effects on a selected track or the entire media project you are working on.

Various developers offer downloadable plug-ins that you will be able to find in the App Store for download. You can also find various plugins on different other producers' platforms that they make available for use.

While sourcing plugins for your tracks, it is important to remember that they won't always be free and often a priced item will offer better sound and production. It is not to say that free versions of plugins are a bad idea, but as with anything, quality takes time and the people who worked so hard to create such an awesome product deserves to get compensation.

The following is a list of free plugins that are extremely useful even for beginners:

1. Ambience by Smart Electronix offers 60 free presets.

2. Mfree FX Bundle has various free presets but even more, can be purchased via the upgraded versions.
3. Synth1 by Daichi Laboratory offers almost 128 presets and is especially useful for someone wanting to work with synthwave sounds. This plugin is one of the best software to use on the market.
4. Vinyl by iZotope offers the user authentic vintage sounds. It is a wonderful lo-fi free software plugin.
5. Vocal Double by iZotope offers a different depth and character to your recorded vocals. It has an efficient workflow that is great for beginners.
6. TDR Nova offers an easy-to-use interface and 4 different dynamic equalizer bands.
7. ValhallaFreqEcho by Valhalla DSP offers the user analog-style echo with a wonderful and easy-to-use knob interface.
8. Tyrell Nexus 6 by u-he is a wonderful synth plugin with an extremely well-planned GUI.
9. Saturation Knob by Soft Tube is one of the best and most useful plugins for any producer wanting to create amazing vocal and instrument sounds.
10. OTT by Xfer Records offers the user a 3-band expansion and compression with this plugin, allowing the user amazing compressor presets.

Mini Tutorial 2

The following tutorial continues onwards from the previous mini-tutorial in chapter 1. You will be testing your knowledge of what chapter 2 had to offer:

1. Open your file named My First Project.
2. In the LCD change the tempo of the track to 94.
3. Select the audio track and change the volume of this track.
4. Select your drummer track and solo it, whilst recording.
5. Mute your audio track and listen to your drummer track.

6. Find a plugin you like and follow the developer instructions to install it into your project.
7. Use the plugin to change the sound you are working with.
8. Use a shortcut to save your file.

This chapter covers content that relates to chapters 3, 4 as well as chapter 5. Any terminology not understood can be found in the glossary just after the introduction of this guide.

RECORDING

You can finally move on to the recording section of the guide. In this section, we will cover how to record your keyboard, guitar, and vocal tracks. While jumping in at the deep end is exciting there are a few things that need to be covered before we can get to the fun recording part.

We have discussed the different types of tracks and what they are used for, this is important as recording your keyboard, piano or guitar sound on an audio track will severely alter the sound that will be produced, and of course vice versa.

Alongside recording, we will also be covering how to record different takes, how to delete and re-use takes and how to record multiple instruments. I will also include overdubbing and the use of a metronome and a tuner inside GarageBand.

While this guide teaches you the basics it is important to remember to do what you are comfortable with and have the ability to do. Jumping into the deep-end and struggling might deter you somewhat, so I suggest taking your time to get to know the software before recording your first track. This will help with any frustrations you might have down the line with any troubleshooting.

Metronomes and Tuners

In this section of chapter 3, I will be covering why and how you use metronomes and tuners within GarageBand as well as outside the software. Both of these things are extremely important when it comes to great sound production and development.

Both of these instruments can be used to make sure that your recording quality and playtime are in sync and using the right tune.

Metronome

The metronome is used to make sure that you are playing in time with any of the other tracks in your project. You can toggle the metronome function on and off inside the software if you wish to keep track of it yourself.

The metronome in GarageBand uses a 'tapping' sound to help you keep track of the time and beats per minute on your project. I suggest always wearing a great quality headphone set when recording. It will allow you to clearly hear any sound that may be off or track any beats that go missing or seem to disappear.

The GarageBand metronome plays in time with the beats per minute or tempo that you chose at the beginning of your project. Once you change your tempo, the metronome tracking will adapt to play at the same rate.

The metronome in GarageBand also has a useful 'count-in' function that allows you to create a short count-in before the recording starts. I find this useful as it gives you a second to hit the record button and get ready to play. This eliminates having to cut out large chunks so scuffling sounds as you move around just before the recording and your playing starts. GarageBand even allows you to select the number of bars that you wish to have the count-in be. You can do this by using the Record menu, and selecting Count-in, here you will be able to select the number of bars from the correct submenu.

You do however not have to use the count-in if you prefer not to use it. It is very much up to personal use and comfort with the software as well as your instrument. You can also change the volume of the metronome. This is extremely useful as having the metronome click too loud might affect your hearing when it comes to recording. Some musicians prefer to only see the ticking instead of hearing it and that is also perfectly fine. You can change the volume by using the GarageBand Preferences menu. This can be found by hovering your mouse over the GarageBand name in the left-hand corner and selecting the Preferences menu. Select the Metronome submenu and adjust the settings to your preferences.

You can also adjust the tone of the metronome. This can be done in the same menu where you adjust the volume of the metronome. The metronome is by default a short sharp sound, by dragging the tone slider left, you create a duller and softer sound while dragging it right will create a deeper sound, almost like two pieces of wood clacking against each other.

You can also toggle the metronome on or off next to the LCD screen. The metronome and the count-in are the two icons just to the right of the LCD screen. The numbered icon is the count in, while the triangular icon is the metronome. When they are toggled on they will appear as two purple buttons.

Tuner

The tuner can be used to tune your instruments but also allows you to stay in tune with software recordings you have made, and any other tracks that you may have already recorded or imported into your project. Unlike the metronome, the tuner is slightly more difficult to control as it has 5 different settings that you need to take into account when using it.

The tuner icon is just to the right of the LCD screen. It is easily identifiable by the pronged design on the button. As with the metronome, toggling the tuner on will make it appear almost like a purple button. When you toggle the tuner on, a tuner pop-up screen will appear to show you the tuning of your current recording. This pop-up will guide you to using the tuner correctly and allow you to record the sound the way you want it to sound.

The tuner pop-up will have a needle display. On the left-hand, it will show the hertz at which the note is playing and on the right-hand, it will show the cents. The needle will swing while you are playing, letting you know when a note is flat or not. If the needle is in the center it means that your note is in tune, while when it swings left, it is flat, and the needle swinging right means that the note is sharp. GarageBand's tuner also has an added bonus where the note that is displayed and the needle will turn green if it is in tune. You can also set a reference for yourself. This lets you select a reference pitch and allows the tuner to guide your tuning a little more precisely. You can do this by clicking on the hertz selection and dragging it higher or lower.

Now that the setup has been done you can play a single note to see if the tuning on your instrument is correct. Play a single note and see what the tuner indicates. If your note is flat or sharp, adjust your instrument until you reach the perfect pitch.

Recording Tracks

The metronome and tuner will help you a lot, especially when you are first starting. It is important to make sure you are in tune and that your time is synced, otherwise you might end up recording hours of tracks that are out of sync or flat and you will have wasted your own time. Once you have set up your metronome, count-in, and tuned your instrument you can move on to the recording section. Each track selection has a different interface when it comes to recording. I will be covering the basics for each interface before moving on to the mini tutorial for his section.

All three track recording sections have an option called Noise Gate, while this is toggled off for audio tracks, it is toggled on for both guitar, piano, and drummer tracks. This keeps any excess noise from moving around the equipment while playing to a minimum, it also allows other softer sounds to be recorded, creating the perfect mix of musical sound and ambiance.

The recording process is much more than just simply hitting the record button and getting a final product. When recording both sound and a software instrument it is important to make sure that you have all the correct adapters and connectors. Each instrument has its own connector and it might be beneficial to invest in a **MIDI** to be able to connect all your instruments with the correct adapters and interface.

Audio Tracks

Audio tracks are recorded somewhat differently than software instruments. This is because vocals are a completely different sound. In this section, I will cover the recording settings for audio tracks meant to record voice or vocals. GarageBand offers a wide range of different vocals, from bright vocals that are considered happy and cheerful to experimental vocals for punk or rock musicians, the software even includes a vocal module for narration. Each vocal module comes with its own set of controls and can be adjusted however you would need them. Controls and their adjustments will be discussed in chapter 6.

The recording block for audio is found underneath the track bar. There are a few settings that can be changed when recording any type of voice audio. The first setting you'll find is the Record Level, this setting changes the volume at which the audio is recorded. If you opt to select the Automatic Level Control, just below the Record Level bar, it will adjust the settings automatically for the entire track.

The second setting that you can change for your recording is selecting the Input. There is a circular icon that has a small dropdown menu to the right hand. The selection in the dropdown menu will be the microphone you have plugged in. There are 3 dropdown items to select from; No Input, 1(Built-in Microphone), or 2(Built-in Microphone). If you click on the single circular icon, it will change to two interlocked circles. The dropdown menu will change. The dropdown menu will only have one selection besides the No Input selection, 1-2(Built-in Microphone).

The third setting change that you can make is the Monitoring selection. If you click the little icon it will switch on monitoring and the icon will become orange. This means that the monitoring has been toggled on. There is also a Feedback Protection tick box that you can select. The feedback protection eliminates any feedback from the recording, this is especially important for Narration or Telephone Vocals as a clear background will make enhancements and effects easier to add and adjust.

Software Instrument Tracks

Software tracks are either guitar, piano, or bass guitar tracks. The GarageBand library has 7 different guitar tracks to choose from. From Clean Guitar, which has its own guitar modules to Experimental Guitar. It also offers you selections from Clean Bass to Experimental Bass. Each of these tracks has its own modules and plugins, all with their own controls and effects.

The Software Instrument track recording settings are very similar to the Audio track recording settings, the only visible difference being the Noise Gate that is toggled on. As mentioned at the beginning of the Recording Tracks sections, the Noise Gate creates a gate to keep sounds below a certain decibel range out of the recording.

If you opt to select no input you will not be able to record any type of sound. This is especially useful if you have a track already recorded that you would like to import into your project. While importing is something you can do, and you can use the library plugins to change the sound, the way that the track was recorded will be extremely important. Importing media will be thoroughly covered in chapter 4. You can add multiple tracks into a project, even recording them at the same time if you can plug in multiple instruments. If you choose to record two separate instruments at the same time make sure that you create two tracks, and link the correct input to each, as the wrong input selection might change the sound quite drastically.

Drummer Tracks

Drummer tracks are the exception to the rule of recording. While the recording section looks similar the entire process is different as you would not need to plug anything in, but instead use the built-in drummer selection and settings from GarageBand itself.

The drummer track selection in GarageBand adds a virtual drummer into any project. Each style of drummer has a unique sound and the playing style can be edited to match another track's style and rhythm. The drummer section does not have its own recording section. This means that you can record the drummer track as you would record all the tracks together, by using the LCD screen at the top.

The editing and changes for the drummer playback can be done in the drummer editor section. The drummer and other track editors will be covered in more detail in chapter 5, but I will broadly explain the type of editing that can be done in the drummer editor in this section of the guide.

The entire layout of the drummer tracks will look different from the audio and software instrument layout. The library will be divided into four different sections. The top two sections are linked to the genre while the bottom two sections cover the virtual drum kit you would like to add to your track.

Before making your drummer selection the software will by default add the SoCal Kyle drummer. You will also see the track-specific and an image of a drum set that is adjustable. If you wish to adjust these settings instead of making a different selection, you are free to do so. Here you can change the Beat Presets, Percussion, Hi-Hat, Kick & Snare as well as the complexity and volume of the drum selection. However selecting a drummer and genre will help you create a more refined sound, and it will only be enhanced by the drum kit you select. Once you have made your selection you will be able to see the Controls and Track panels beneath the Track Header section, and the image of the drum set will be replaced.

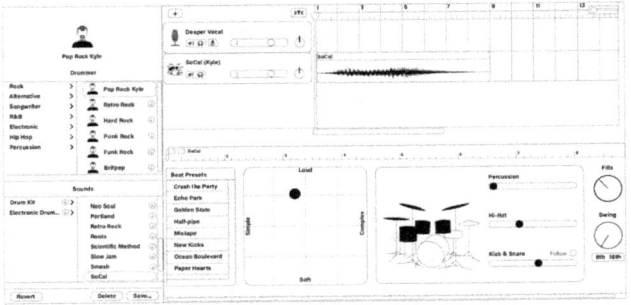

Mini Tutorial 3

In this mini-tutorial, the questions will cover everything that has been discussed in this section:

1. Record a voice section on your audio track, mute your second and third track.
2. Record your entire track by unmuting the software instrument and the audio track.
3. Make sure to use the tuner to tune your instrument before moving on to the next step.
4. Record a new track and use both the metronome and count-in.
5. Unmute all your tracks and record them all together.
6. Use the shortcuts to save your tracks between recording phases, as well as after you are finished.

This chapter relates to content that will be covered in Chapters 2, 4, 5, and Chapter 8. Please feel free to move to those chapters for more in-depth content surrounding the aforementioned content.

MEDIA

In this section, we will be covering media, and how importing any media into your project will help you elevate the sound and quality of your production. I will also cover how to use loops and how to import the video footage to create a film.

The loops section in this chapter will cover all the loops that GarageBand has to offer, this section will also cover how to use loops in your tracks. The media section of this chapter will cover different media and how to import such media as well as how best to utilize the features in GarageBand.

Apple Loops

Apple Loops are pre-recorded sections that can be added to any project to allow you as the user to build the perfect base for any track, whether you're working on a narration project or adding sound to a film or a blog. Apple loops can both be repeated and stretched to fill the allotted space for any music or sound.

Loops can either be found in the GarageBand library and

GarageBand automatically creates a region for you when you add a loop to your track. When playing the project the regions will automatically play at the project's key and the project's tempo. Note that no matter how many loops you use, even if they have different keys and tempos then they will change to match the project key and tempo. There are 3 different types of loops found in the Apple Loops library. As with most things in GarageBand each category is color-coded, the same color as the track selection they are from.

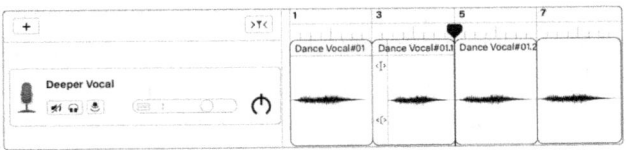

Audio Loops (Blue)

Loops that are blue are audio loops that can only be used with the audio track selection. These loops can be edited and changed in the Audio editor beneath the tracks section.

Software Loops (Green)

Loops that are green are considered software instrument loops. These loops can only be used on a software instrument track. They will however be edited in the Piano Roll and Score Editor, also found beneath the tracks section. GarageBand also allows you to convert software instrument loops to audio loops by adding them to a selected audio track.

Drummer Loops (Yellow)

Loops that are yellow are considered Drummer loops and host all the necessary settings and information that is needed to play a drummer track. You are also able to change the Drummer selection for your drummer track in the library or in the Drummer editor. Like the software loops, you can convert drummer loops to audio loops, however, you have to convert them to software instrument loops before converting them to audio loops and adding them to your audio track.

The loops library that is offered by GarageBand is vast and has various different sounds, genres, and moods. Along with this expansive library, GarageBand offers you the ability to create your own loop selections by creating a favorites tab as well as loop customization and loop creation from scratch. The following sections in this chapter will cover these concepts and their respective explanations.

Loop Customization

Loops are small sections of sounds that can be repeated over and over in each track. In this section of the chapter, I will cover the basics of using loops in your tracks as well as creating your own loops.

GarageBand Basics 41

Searching for Apple Loops

Loops are found in the Apple Loops section, in the Loop Browser. It is not found in the library section to the left where you would find your track selections. To find the Loop Browser, you can click on the loop icon in the top, far-right area of the menu bar, or use the view menu selection at the top and click on the 'Show Apple Loops' menu item. This will enable the Loop browser view.

The loop browser has different methods of viewing all the loops that GarageBand has to offer. The default view that will show when opening the Loop Browser for the first time is the buttons view, which will look like a grouping of buttons with names of genres on them. The easiest of these methods is to view the loop packs. Loop packs are different genres. To access the loop packs, simply click on the selection at the top next to the Loop Browser table heading. A small pop-up menu will appear with all the different packs that you can choose from. Select the pack most applicable to you and find the loops you would like to use.

You can also search for loops, by name or instrument. To do this you will be using the search functionality that is included in Garage-Band. Type any keyword you need our pack or loop to have into the search field and press the enter key. This will return every file that GarageBand has with that keyword in its name or title.

You can also filter the type of loops you would like depending on the loop type. You will do this in the loop browser section. Just above all the loops, you will find an icon that is three colored lines, blue, green, and yellow. This signifies the different loops. Clicking on this icon will create a pop-up menu that has four options: All Loops, Audio Loops, MIDI Loops, and Drummer Loops. Unticking any of these boxes will exclude that loop set from your search results.

Restricting loop types can also help you find the correct loops when using the filtering options for finding a specific loop. You can restrict loops by package, scale, and by type. By restricting the package selection, you will actively exclude all electric guitars for example. Scale restriction will restrict all loops with a Minor scale and type restriction will exclude all loop types that are unticked in the pop-up dialogue mentioned in the filter option.

You can also view loops without having to add them to your tracks. Any loop that is previewed will be played in tune and tempo of your project. It will also play alongside your project, even though it has not been added to your project. You can use the volume slider at the bottom of the loop browser to lower the volume of the selected loop. To play and pause the loop, simply click on it.

Favorites in Apple Loops

Setting your favorite loops in GarageBand will make it easier to find the sounds you love without having to filter or try and remember that one word in the title when you have to use the search functionality. Here's how you can set up loop favorites in GarageBand.

In the Loop Browser window, on the far right, there is a small column with a heart icon at the top. The Heart icon represents your favorites. Next to each loop, there will be a tick box underneath the heart icon. If you tick the box, the loop will be added to your favorites list. You can access these favorites by selecting the favorites button in the button view that the loop browser automatically opens up in or the favorites in the column view.

Once you have searched and favorited the loops you wish to use now, and in future projects, you can finally add them into the appropriate tracks. Once they have been moved into the tracks section they can be edited just like tracks can. They can be shortened or stretched or even edited by the editors.

Some of the loops you find will be part of a loop family. This means that they are from the same genre, but have different types of sounds. These loops will have a number sequence at the end of the name, usually starting with a 0 and ending in another numerical digit like 3 or 6, depending on how many loops there are in the family.

There are various ways to add the loops you want to use to your tracks. The easiest is to just drag it to the track where you want to use it. If you have not added a track yet, you can still add a loop to the area where the tracks are located as GarageBand will automatically add a track that's the same type where the loop can be placed.

If you want to use a software instrument (MIDI) loop to a drummer track, or a drummer loop to an audio track you can simply drag it to the track you want it to be a part of and GarageBand will automatically convert it to the track type. This cannot be undone unless you delete and add a new loop or track.

Custom Apple Loops

Creating your own custom loops is an amazing way to truly create a unique sound. GarageBand offers the capability to create your own loops and save them, for future use and in your favorites tab. Apple loops are generally created through regions instead of using full tracks.

The following steps can be taken to create your own custom loop in GarageBand:

1. In your GarageBand, add any track of your choosing, as well as a region.
2. Open the Loop Browser section.
3. Drag your region from the track selection into the Loop Browser.
4. A dialogue will pop up with the name "Add Region to Apple Loops Library."
5. In this dialogue box you will do the following:

   - Change the name of the loop to whatever you want to call the loop.
   - Leave the type selection as "Loop."
   - Select the scale for your loop from the dropdown menu.
   - Choose a genre from the dropdown menu.
   - Choose a key for your loop from the dropdown menu.
   - Select an instrument from the "Instrument Descriptors"

section, and then select an instrument from the right-hand column of the box.
- Select one of the mood buttons on the right-hand side of the column.

6. Click the Create button when you have done all of these selections and you're done creating your own loop.

These loops are automatically added to the Loop Library and Browser and can be found when searching for the name or when filtering for the specific loop type that you assigned to the loop. You can also add it to your favorites for easy finding the next time you want to use it.

Loops can also be found outside of GarageBand or creating your own. There are various places online where you can buy Loops from a third party, or even just download them for free. You can easily add these third-party loops to your loop browser and use them in future projects. Simply drag your third-party loops from the finder window, into your Loop Browser window and GarageBand will automatically index the loops in the Loops Library.

Media

Adding any type of media to your project will drastically enhance and better your project. In this section, I will cover all the media you can add to your project and how to do so. I will also cover how to add film footage to your project to create a music video or a short film.

GarageBand supports 6 different file types when it comes to Audio imports. Those file types are as follows:

- AIFF
- CAF
- WAV
- ACC (ACC protected files are excluded here)
- Apple Lossless

- MP3

While these are the 6 main audio imports that can be added to a GarageBand file, you can also add MIDI files into your project. While your MIDI files do not contain the actual recordings of your sound, it does contain all the necessary data that allows GarageBand to create a sound with the software instruments offered.

Audio files can be imported into GarageBand in two different ways. Through the Finder, through a Music app, I will also cover how to import a MIDI file. Each of these methods is pretty basic and easy to follow.

To import a file from your Finder into GarageBand you simply open both the Finder window where you have saved your audio file and GarageBand and you drag it into the tracks area. GarageBand will automatically create a new track for you.

To import an audio file from a Music app is similar. Open the Music app you want to use the track from, open your GarageBand, and drag the track from the music app to GarageBand, where it will automatically be made into a track.

With both of the above-mentioned methods, be sure to drag the track below any tracks in the track area, as it might delete and replace the spot to where you dragged it.

Importing MIDI files from Finder is somewhat similar. You have to create a software instrument track before adding the MIDI file. Open the Finder window where you have saved your MIDI file, as well as GarageBand where you have saved the software instrument track. Drag the MIDI file into GarageBand, if it appears in multiple software instrument tracks, simply select the one you wish to use and delete the other.

Adding Film Footage

GarageBand offers producers much more than just instrument recording, it also offers you the ability to add sound, narration, or replacement audio to a movie file. While GarageBand does not allow you to edit the actual movie footage, it does allow you creative freedom when it comes to any audio changes you want to make. GarageBand does restrict the movie file to 1 per project.

The best option would be to play the film inside a Movie window that happens to be floating over your GarageBand window. When playing the track, the film will automatically play frame by frame along with the track. The floating window as with any other window can be resized to whichever size you would like it to be.

To open a film inside GarageBand is rather straightforward. Click on the File menu option and when the pop-up menu appears, select the "Open Movie" selection. This will open the movie track in the track section, to be able to add a movie into the track, click on the word 'movie' and then select "Open Movie" again. This will bring up the Finder window where you can make the selection of the film you would like to add to your project.

Once the movie has loaded you will see that GarageBand has automatically added a track beneath the movie. This track is the film's audio track, this will be here you will be making any changes to the audio and sound. I advise against moving the audio track away from the movie as this will cause issues with the synchronization of audio to film later on. To move around within the movie, and the audio, you can either use the slider located at the bottom of the movie window, or if you prefer to use a more precise approach, opt for the transport buttons in the playhead.

If you selected the wrong film and would like to replace it, you can simply drag a new movie file into the project, or use the same steps as above. If however, you want to remove the film entirely I would suggest clicking on the "File" menu option, and once the pop-up menu opens select movie and then remove movie, or click on the word 'Movie,' in the track section and select the "Remove Movie" option, this will remove both the movie and the track along with it.

Above the movie track, you will see small frames of each scene. All the frames will be aligned to the left of the track, except for the final scene which will be right-aligned. The left alignment margin is

GarageBand Basics

synced to the correct frame and time position. This alignment enables you to see both the first and last scene, no matter how much you have zoomed into or out of the track and film. To be able to view this, if it does not show on your screen, select the "Track" menu and select "Show Video Track."

The creation process for a sound file on a film is the exact same as when you are working with any other music production in GarageBand, you can record your own sounds, add narration, use regions and loops. The important thing to remember however is to always make sure that your sound and movie files sync up. The best way to achieve this is to select the small drop-down arrow in the LCD screen, next to the Scale selection and instead of showing 'Beats & Project' select the 'Time' option, you can also adjust the tempo here if some beats and key points are not aligning.

When you are finished changing the music and adding effects and you are completely finished with the project it is important to remember that unless you muted the Movie Audio track, the original sound will still play. Also, make sure that you export the film correctly to ensure that your audio does not disconnect or become out of sync.

The best way to do this is by using the "File" option. Once you have selected the file, select "Movie" and then click on "Export Audio to Movie." This will ensure that no files are corrupted or end up causing errors later on. Once you have selected the export menu, an Export dialogue will pop up where you can select the export quality. Each quality setting has a small explanation below to help you make the best selection for your project. Once this has been done, give your film a name, select where you wish to save the file, and press the enter key. This will save the film in the location you selected.

It is important to note that if your audio is longer than the film, GarageBand will automatically lengthen the film to match the audio track, so be sure to check the length of your track before exporting your film.

Mini Tutorial 4

In this mini-tutorial, we will be covering all the content regarding loops as well as adding media to your projects:

1. Open your project.
2. Add a loop of your choosing for each track that you have added to your project.
3. Use the shortcut to save your project with the name 'I've gotten this far.'
4. Create a new loop of your own or install a third-party Apple loop.
5. Save your project.
6. Add a media import to your project, it can be an audio file or a film file.
7. If you opt for a film file, edit the track to match your film and export it.
8. If you opt for the audio file, complete your track and export it.
9. • Remember to save!

The following content is linked to Chapters 2, 3, and 8. They cover topics related to this chapter. Feel free to move to those chapters for more information.

ARRANGEMENTS

Arrangements are made within the tracks area. They are the building blocks of a great track. In this chapter, I will cover how to use regions to create the best arrangements for your needs. This chapter will also cover the editors used for each track and music notation that can be created in GarageBand using the notes section.

Knowing how to navigate the track area will help you greatly when creating your arrangements. It is also important which settings to include and toggle on or off to get the best possible experience within GarageBand.

Regions

Regions are the small blocks of recordings, loops, or sounds that are used to create your track. Every single loop, track, or recording is called a region. The entire block of a track, that includes all recordings, loops, and segments is called the track.

Regions can be edited and changed in the following ways when they are located in the tracks area. Each of these edits will be discussed:

Selection of Single or Multiple Regions

Most edits for regions must be done while the track is selected, to do so simply click on the region, and it will be highlighted, signaling that it has been selected. You can select multiple regions by holding down the shift key on your keyboard. If you want to select all the regions in a track, simply click on the track header, or if you want to select all the regions in a project simply use the Select All shortcut, (Command + A).

Cutting, Copying, or Pasting Regions

Regions can be cut, copied, and pasted in GarageBand. It is a great functionality as it allows you to copy various sounds for re-use everywhere in your project without having to constantly recreate the sound. To copy, paste or cut a region, simply use the shortcuts provided in GarageBand. Copy (Command + C), Cut (Command + X) and Paste (Command + V). Once you have copied the region you can move the playhead to where you want to paste the region and simply use the shortcut to paste it in that section.

Moving Regions from One Spot in a Track to Another Spot

Regions can be moved from one section in your track to a different section, or even from one track to another. To move the region from one space in the track to another, simply click and drag it left and right in the track, this will move the region to a different area in the track. If you want to move a region to a different track, the same logic applies, simply click on the region and drag it down or up into the track you want to add it to.

Looping Regions

Looping regions allows you to repeat the sound or section of that region for a few seconds or minutes depending on your project. To loop a region, hover your mouse over the top right-hand corner of the section until it becomes a circular icon, then simply click and drag it to the small line it will show to create the loop. You can also loop a region to play repeatedly. Click on the region while holding the control button down and select the menu option Loop On/Off, this will loop your region repeatedly. To turn off the loop simply control and click the region again and make the same menu selection.

Resizing Regions for Longer or Shorter Playtimes

Resizing regions allows you to either stretch or shrink the sounds so they can play for a longer or shorter period of time. To resize the region, hover your mouse over the lower right-hand corner of the region until an icon with two arrowheads and a bracket appears, click and drag the region to be shorter or longer. You can also resize adjacent regions, to do so, hover the mouse over the area where they meet and when the icon appears, simply click and drag the regions, adjusting both. If there is a gap between the two regions, the icon will appear over the upper-right corner of the first region to the left.

Joining and Splitting Regions

Splitting regions can be done inside the track area. If you split a region that has been copied and used elsewhere, only the selected region will be split. To split a region, select the region and place the playhead where you want the region to be split. Hold the control button and click, a pop-up menu will appear, select 'Split at Playhead' and the region will be split. Only the selected region will be split, even if there are other regions in a track below or above the region you have selected.

To join regions is just as simple. Regions have to be adjacent to each other and they must be on the same track to be able to be joined together. To join the regions together, press shift and click on the regions you would like to join, then select the Edit menu and select 'Join Regions,' this will join the regions. The region will be renamed to the first region's name. GarageBand creates an entirely new region instead of just joining the regions.

Creation and Renaming of Regions

GarageBand offers you the functionality of creating regions, outside of using the library regions and Apple loops. You can create 3 different types of regions; empty MIDI regions, audio regions from the audio in the tracks area, and drummer regions. Each of these has its own way of being created.

Empty MIDI regions are created by pressing the command key and clicking on a software instrument track where you want the empty MIDI region to start.

An audio region from an audio file can be created by pressing the command key and clicking on the audio track where you want the region to start. Once you have done this a pop-up menu will appear, simply select a file from the open menu.

Creating a drummer region is just as simple as the other methods. Simply press the command key and click on the drummer track where you want the region to start.

Keeping your tracks and project organized will help you in the long run, and one of the easiest ways to do that is by renaming each track to match the output it is creating. Having 12 vocal tracks will make you crazy and being able to just quickly filter through 3 female vocals is much easier. To rename a region simply press the control key and click on the region, select 'Rename Region' from the menu and type the new name for your region, once this is done you can hit enter and your region will have a new name.

Deletion of Regions

With GarageBand, you can delete one or multiple regions at a time. You can also move the regions that follow the deleted region up to keep from having blank spaces left open in your project.

To delete an app without moving the rest of the content, you can simply use the delete key. Select the region you wish to delete and press the backspace button. This will delete the region without moving any of the other regions up.

To move up the other regions after you have deleted your region you will have to do something a little different from just hitting the delete key. Select the region(s) you would like to delete, select edit and choose the 'Delete and Move' menu option. This will delete the selected sections and move the other sections forward into the space that has been left open.

Ruler

GarageBand allows you to use the ruler functionality to align your regions to beats, to the grid, or even snapping them to other sections, or items within the track area. Regions can also be aligned with other tracks, to play only when a selected loop from a different track is played.

The ruler in GarageBand is horizontally on top of the Tracks area. The ruler changes depending on the grid selection that can be made. When using the musical grid, the ruler will show the time formatting in bars, or in 8-second intervals. If you opt to not use the musical grid, the ruler will show the time formatting in minutes and seconds.

The ruler functionality is a default setting in GarageBand. It is extremely efficient and allows you to move regions, loops, or recordings to a specific timestamp if you need to. The playhead, as previously mentioned, will highlight the section that is currently playing. You can also use the ruler section to place the playhead in the exact right spot when editing.

Alignment Guides

Alignment guides will appear when you move around loops, or sections of recordings, mainly to just help you correctly align the track when attaching them to a different track or loop. They also appear when one region is touching another. These are indicated by the yellow lines that appear around the selection that you are moving around.

Alignment guides can be turned on or off in the editor menu. When turning alignment on it automatically turns the snap to grid on as well. This means that the moment two sections align, they will snap to the alignment grid.

If the alignment guides and the snapping to the grid are turned on, and you perform actions such as moving regions, moving the playhead, changing automation points, or moving loops into the track area, your item will move to the next block on the grid. Grid value and size are determined by how far you have zoomed into the project.

The playhead, the line and arrow icon at the start of the tracks, extends from the very first track all the way to the final one. It will move from the left to the right as the play-through is happening. You can use the icon at the top to drag and move where the track should start playing while you are working on a specific section, this allows you to alter and change any loop or recording without having to listen to the entire track from the start. This will save you time, especially if you are working on a particularly long project.

Arrangement markers can also be added to regions and to allow you to better manage your entire project. Arrangement markers can also be mentioned in the notation section, to help you remember the structure or to make it easier for a team to work on the project.

Editors

Editors are a way to change and alter each track far more individually than you would by altering the master track. While the editor is a powerful tool it is important to remember that the editing capabilities are dependent on the track selection that you have made.

Editors are a way to change and alter each track far more individually than you would by altering the master track. While the editor is a powerful tool it is important to remember that the editing capabilities are dependent on the track selection that you have made. There are four regions that can be edited in the editor sections. These regions are all edited in a different editor.

In the following section, I will be covering the basics of using the 4 editors that come included in the GarageBand software. The first three will be covered broadly in terms of settings and changes whilst the final editor will be covered in more detail.

Remember that the editors are a way for you to refine the sound you want to create. To alter and edit the tune and pitch of a specific region, or to add more effects or changes to the master track.

Each editor can be resized inside the GarageBand window. You can do this by dragging the bar above the editor up, enlarging the view of the editor you have. The same can be done to make the window smaller, only by dragging the bar down. This will shrink the editor window and allow you more space to see the track area or the library. You can also zoom into the editor if you need to. Garage-

Band offers a zoom slider in the right-hand corner of the editor window that allows you to easily zoom in and out of the editor pane, allowing you more precision and control.

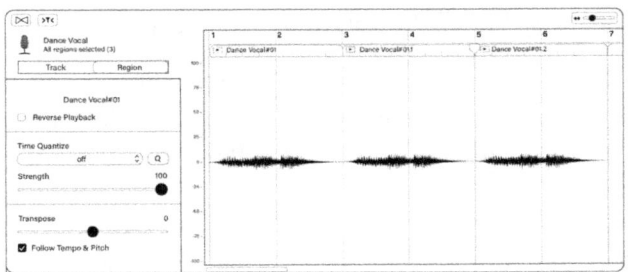

Audio Editor

The audio editor allows you to do various edits to audio regions. These edits include, but are not limited to:

- Spitting and joining regions
- Renaming and transposing regions
- Following the tempo
- Volume control
- Enhanced timing
- Adjust audio region tuning
- Edit audio files that have been converted to audio regions

The audio editor allows you the freedom to entirely change and alter any audio track inside your project. Zooming in will allow you a better view of the waveforms in the tracks, this means that you have more control over splits and cuts in the editor itself.

I will cover how to do the basic adjustments to audio regions in the following section. These will only be broad detailed explanations. It is important to remember that the changes that are made, are always non-destructive, which allows you to go back to the original version of this region or file if you so wish.

If the editor is not automatically opened when you have a track selected you can open the editor simply by double-clicking on the audio track. You can resize the editor vertically and horizontally by dragging the bars to the left and top of the editor. You can close the editor by clicking the edit button in the control bar.

The changes that can be made will change depending on if you have selected one track or multiple regions. If you have selected one track and no individual region, the following information will be shown in the editor:

- Pitch Correction: Finds the pitch correction that needs to be applied to regions for perfect pitch
- Limit to Key: Limits the pitch correction to notes that are only in this project
- Enable Flex: Allows flex editing to be done for the selected track

The menu looks somewhat different when you have selected more than one region:

- Region Name: Allows you to edit the region's name in the editor
- Reverse Playback: Plays the selected audio in reverse
- Time Quantize (pop-up menu - Q): Quantize the note timing and any other events in the selected regions
- Transpose: Allows for change on note pitch in the selected regions
- Follow Tempo and Pitch: Allows you to adjust if the tempo and key stay the same as the project's tempo and key

These are not the only settings that you can find inside the editor, you can also do the following edits to tracks and regions inside the editor. These edits do not have a separate menu as they are mainly done through menu options:

- Cut, copy and paste regions

To cut a region in the editor, simply select the region, select edit, and click on the 'Cut' option.

To copy a region, select the region, select edit, and then click on the 'Copy' menu option.

To paste a region in the editor, put the playhead in the area where you want the region to be pasted, select the edit menu and click on the 'Paste' option.

- Move, split and trim regions

Moving a region inside the editor is similar to when you are doing so in the track area. Click and drag the region to where you want it to be.

When splitting a region in the editor, place your mouse on the lower half of the region until the pointer changes to a marquee pointer. Drag the marquee to where you want to split the region. The marquee tool will split the region for you. The part that is outlined is the section that can be deleted if there is no use for it.

To trim a region in the editor you have to place your mouse pointer over the lower-left edge of the region until the trim pointer pops up. The trim pointer is two arrow icons with a bracket in the middle. Drag the pointer to the length you want the region to be and GarageBand will automatically trim it for you.

The audio editor also allows you to quantize the tracks and/or regions in the editor. Quantizing simply means automatically correcting the timing of your tracks or regions. You can do so by using the pop-up menu (called Q) or the strength slider in the editor. By selecting a note value in the Q menu the software automatically quantizes the track or region to that note.

You can also alter the pitch and correct it by using the pitch slider. If you want the pitch correction only to apply to notes that are in the project key, tick the 'Limit to Key' tickbox. You can also edit notes and beat timing in the editor. By selecting the Flex Time option,

GarageBand will simplify the process of note and beat timing. Flex Time allows you to stretch and move the notes and timing without having to move, trim or cut sections out of your hard work. You use Flex Time by placing markers on the track.

To place Flex markers on your track, select the Enable Flex tickbox, then click on the waveform section you want to edit, and the Flex Time markers will appear. The flex markers will appear right where you click and at the preceding and following fading of waveforms. To move these markers simply hold down the Option key and drag them to where you want them to be. When they have been set up, drag the flex marker to the right or left.

Piano Roll Editor

The Roll Piano editor is the place where all software instrument editing will be done. Most of the MIDI editing is done inside the piano roll editor. Software instrument editing is also done in the piano roll editor.

The following edits are done inside the RPE:

- MIDI region editing
- Transposing MIDI regions
- Enhancing timing
- Note editing
- Edit MIDI regions in the Score editor

The piano roll editor is mainly used to help the editing and changing of MIDI regions and recordings. It can however also be used to help with the editing of software instrument tracks. Horizontal lines on the editor will indicate the time position of each note whereas the vertical lines will indicate the pitch for each note. A keyboard is also seen on the left-hand side, to allow you to reference any notes that you may have misperceived or need to change.

At the top of the editor, there will be two buttons, one named 'Piano Roll' and the other named 'Score.' The Score editor will be covered later in this chapter. As with the other editors, you can also size the

GarageBand Basics

roll piano editor vertically and horizontally. Like with the Audio editor, the settings that can be changed differ when choosing regions versus choosing tracks:

- Region Name: Edit the name of the region or track that you have selected
- Time Quantize (pop-up menu - Q): Quantize the timing of notes
- Transpose: Changes the pitch of the notes in the selected regions or tracks

When individual notes are selected the following menu will be on the left-hand side of the editor:

- Insert (pop-up menu): The note value selection that you wish to insert
- Time Quantize (pop-up menu - Q): Quantize the timing of notes
- Velocity: Changes the value of the velocity of the selected notes

You can add, edit, copy and move notes into the track or region that you have selected in the roll piano editor. To add notes to the selection you have made, press the Command key and click on the section where you want to add the note. Select the note you want to add beforehand. The note will be added where you have clicked in the track area. To edit a note, simply click on it. It might be easier to do so by zooming into the section so notes appear bigger. To select multiple notes at a time hold down the shift key while clicking.

To select notes of the same pitch, simply click on the corresponding note on the keyboard to the left of the editor. This will select all those specific notes. You can also select all the notes of multiple ranges simply by holding down the shift key and clicking on the various notes on the keyboard.

The roll piano editor also allows you to move notes around inside the editing area. To move a note, select it and drag it either to the left or the right. Oftentimes moving notes like that will be difficult as the increments are too large, as the notes will snap to the grid. To overcome this you have two options, you can either hold down the control and shift keys while you drag the note to where you want it to be or just hold down the control button. While the latter is easier it is important to remember that you will need to zoom in before doing so. Also remember that you can use the shortcuts for copying, pasting, and cutting any notes in the piano roll editor.

You can also opt to change the pitch of one or multiple notes in the track selection you made. To change the pitch simply drag the notes up or down in the editor. You can also resize notes like you would with regions. Simply drag the left or right side of the note horizontally and it will extend or shorten the note. You can resize multiple notes by holding shifts and selecting the notes and then dragging their edges left or right to resize.

You can also use the editor to change the velocity of single or multiple notes. The velocity slider is in the left editor panel, it indicates how hard the note was struck. You can use the slider to change the velocity as it will change the note volume. Deleting notes can be done by selecting the note and hitting the backspace or delete button.

As with the audio editor, you can also quantize the notes in the roll piano editor. Select the notes or the regions you want to quantize and use the pop-up menu to select the note value you want to use for quantization. The slider will only be accessible if you have chosen a MIDI region. Quantization will automatically happen for note selections.

Drummer Editor

The drummer editor is used to enhance the drummer track. The drummer tracks are virtual drummer tracks that have been pre-recorded. There is a vast library filled with various tracks that can be chosen. The drummer editor allows you the ultimate control over how your drummer track sounds. The following sections can be edited and changed inside the drummer editor:

- Drum genre
- Drummer region settings
- Preset changes

While all of these settings are also chosen in the library when selecting the best drummer for your project it is important to remember that small changes in the editors can create large differences in the sound production. The drummer editor is mostly used to enhance the sound that you get from your virtual drummer. This is discussed in Chapter 3.

Music Notation

Music notation is generally used within the Score editor. The score editor allows you to see the notes, rests, and any other information that is given through the MIDI regions and plug-in data. You can also use the score editor to export the score to be able to print it out.

The score editor allows you to edit the notes in a more musical form. It also allows you to add notes to your score. By using the Command and click method you can add notes to specific sections. The score editor also allows you to make additional edits to your notes. The following edits can be made in the score editor, and will be translated into the actual track or region you made the edits on:

Note pitch changes

To change the pitch of a single simply click on the note and drag it up to raise the pitch or drag it down to lower the pitch. Lowering the pitch of notes is especially useful if you want to re-use a melody or chorus without repeating the same sound.

Note length changes

Note length changes can be accomplished by dragging the edges of the note to the left to shorten it or to the right to lengthen it.

Note velocity changes

Velocity changes directly impact the volume of notes. To change the velocity of each note, simply click the note and drag the velocity slider, in the editor pane, to the left to lower the velocity or to the right to raise the velocity.

Note quantization

Quantization means automatically adjusting the timing of notes. The process is simple and done exactly the way you would do it in the audio or piano roll editor. You select the note or section you wish to quantize, select the note to which you would the quantization to happen in the drop-down menu, and hit the enter key. It is important to Nate that only MIDI notes or MIDi regions can be quantized in the score editor.

Along with these changes, you can also make basic changes like moving, cutting, copying, and deleting individual or multiple notes. In the score editor, you can also rename the regions you are working with, simply by clicking the region button in the score editor. It will open the region score editor pane, and you will be able to alter the name of one, or multiple regions.

GarageBand also offers you the capacity to print your music notation, either for future use or for other recording purposes. When printing your music notation, it will include all the notes and even any pedal notations that you might have made. Before printing make sure that the score editor window is active and that the notation is visible for whichever track you want to print and simply use the shortcut to print (Command + P).

The music notation will be printed in the standard letter size with the project title at the top of the page. The tempo will be shown in the upper-left section of the page, and the composer's name in the upper-right section. You can change the composer name in the 'My Info' preferences tab for GarageBand. The instrument name that was used in the track will be shown before the bar and the number of the very first measure will be located on the left edge. Page numbers are shown in the center bottom of the page.

Mini Tutorial 5

The tutorial in this section will cover all the content covered in Chapter 5, as well as knowledge acquired from previous chapters:

1. Using shortcuts open the 'I've gotten this far' file.
2. Create your own arrangement.
3. Use any of the editors to help you fine-tune your arrangement.
4. Do some note editing in the Score Editor.
5. Remember to save your files.

Bonus Question: Print out your music notation.

Please note that some of the information in this chapter overlaps in Chapter 7.

CONTROLS

This chapter will cover mixing, and how automations and smart controls will help you create a better sound. GarageBand offers a myriad of control and automation options that will help you refine your sound to be the perfect blend of melody, chorus, and base.

Automations are a basic way of production where mixing tracks and creation of the best type of sounds is made. I will briefly cover automations and how to use them in your project before delving a little deeper into smart controls, amps, and pedals that can be used within GarageBand.

Mixing

Mixing is considered the method used to create a cohesive sound, by blending all your tracks and regions together to create a dynamic sound.

Mixing can be used to alter any of the tracks in your project, including the master track. Mixing does not just include automations but also includes adding effects to the track to create the perfect sound.

Producers will generally follow a few steps when they go through their mixing process. The process of mixing is done in steps to stop any destructive editing from happening and any weird sounds or clicks to be eliminated early on. I suggest following the same steps to help you eliminate any sounds or tracks and to create a more cohesive dynamic sound.

The mixing process is constructed and executed in the following steps:

1. Set Volume

Setting the volume so that each instrument and element can be heard clearly is important. It will allow you to let the most important part of your track be the focus.

2. Set Pan Positions

Instruments should be placed inside a stereo field. The pan positions play an important role as you want the most important tracks and instruments to be placed in the center so that the music is equally distributed to both the left and right headphones or speakers. Any other tracks, normally those that have unusual sound or effects, should be placed on the far sides to stop them from overshadowing the main track.

3. Effects

Effects can be used for various reasons. Effects like compression are generally used to highlight certain sounds in your tracks, while Reverb is used to create a different space. Effects offer a wide variety of changes and adding them in will blend your tracks together far more than you think. It is important to use the right effects for what you wish to accomplish. Play around with some effects to see how they impact your tracks and overall sound.

4. Automation Curves

Automation curves are used to create different changes over time. Automation curves are typically used to create more dramatic changes to sounds but they can also be used to highlight certain sections of sound at a specific point in time. These changes are often changes to tempo and pitch.

It is important to remember that while you are mixing you can enhance and accentuate certain sections of sound by muting tracks or regions or even soloing specific tracks.

Automations

As mentioned before, automations are a way to enhance the tracks you have already created over a period of time. In this section, I will cover how automation curves work and how best to apply them to your project.

Automation curves are identified by the curves and points that will be created on the track of your selection. They will also be a similar color to the track selection, green being instrument tracks, blue being audio tracks, and yellow being drummer tracks.

To be able to add automation points to your tracks you will need to open the track's automation curve. To open the automation curves for a track, simply press the A key. The automation curve will show up and the track header panel to the left of the waveform will change, showing all the icons and menus you will need to be able to add automation curves.

The automation button will be located right next to the headphones icon in the track header. A drop-down menu will also appear, allowing you to select the desired effect or automation you would like to add to the track. To be able to use the automation curve you will have to press the automation icon. Any pre-existing automations on the track will be shown once you hit the button.

Once the automation has been toggled on, you can use the drop-down menu to select the parameter, or item, that you would like to automate. The automation menu offers 4 different selections:

1. Volume
2. Pan
3. Echo
4. Reverb

The volume and pan menu items can be adjusted in the track header, but if you wish to use the automation curve for a more refined volume fade or pan change you can make the selection in the drop-down menu.

Once you have opened the automation curve, adding automation points is as easy as clicking on the section where you want the automation to be. GarageBand will automatically add the automation point where you clicked. You can move them left or right, after they have been placed, to a new time section, or up and down to create a new value for that automation point.

You can select multiple automation points by dragging a small box around them, or holding down your Shift key while clicking on them. You can also copy automation points and move them to a different section in your track, to do this, click on the automation points and hold down the Option (Alt) key while dragging them to the spot where you want them copied to.

You can move or copy entire regions with their specific automations by going into the Mix menu and selecting the Move Automations with Regions option. To delete an automation point you can double-click on the point and it will be cleared from the automation curve. You can delete multiple automation points by selecting multiple points and hitting the delete key. Make sure that you have selected no regions as those will be deleted if you have added them into the selection as well.

Smart Controls

Smart controls are created to help you control the sound that instruments make. These controls are built into GarageBand. You can also import your own or download some amazing third-party plug-in controls that are compatible with GarageBand. Smart controls were designed to change both software instrument sounds and effects.

There are different smart controls, and each of them has its own set of control systems. The screen controls are typically used to change the sound. Each screen control is labeled to let the user know how their adjustment will change the sound of the project. Smart controls include the following control edits that can be made; EQ, reverb, track-specific changes, and tonal control. Each track type will have its own controls that can be manipulated so be sure to check that you have the right track selected when making control changes.

In the following section, I will cover the effects that GarageBand offers in the smart control panels. The following effects are considered the 7 main effects:

1. Compressor: Compressor effects even out the volume for a smoother sound.
2. Delay: Delay effects repeat the selected sound to create an echoing sound.
3. Distortion: Distortion effects change the tone of tracks.
4. EQ: Equalization effects change subtle and dramatic changes.
5. Modulation: Modulation effects repeat sounds but modulate the repeating section.
6. Noise Gate: Noise gate effect reduces the lower-level sounds in the track.
7. Reverb: Reverb effects create different room and environment sounds within the track.

Besides these 7 effects, there are also other effects that are instrument specific. Effects like filters, pitch shifters, and amp models.

There are various types of smart controls, the following is a list of all the smart controls you can find in GarageBand. Each of these smart controls has its own controls that can be adjusted and edited to your liking:

1. Guitar and Bass

Electric guitar controls are synonymous with tonal and gain control while acoustic guitar controls are linked to high and low controls that add body to the sounds you wish to change. Effects like Ambience and Reverbs can be used here to adjust the spacing size and adjust the reverb between patches.

2. Drum Kits

All the drum kits offer you multiple control selections to change every single sound the kits make.

Acoustic kits offer you changes like compression, room, and tone while the electric kit offers you controls like Cush and Drive, Reverb, and Low and High Cut filtering.

3. Orchestral and Mallet

Most of the controls for the orchestral instruments are Low and High controls that change the sound and cut off the top and bottom of sounds to create a cleaner track. Various effects like Chorus, Reverb, and Delay allow you to create spacing and reverb length changes within the track.

Some patches may also include tonal controls to change the sounds of specific instruments. Instruments that are considered brass or string instruments often have controls linked to how the instrument is played, enabling sounds like staccato modes.

4. Vintage Electric Piano and Piano

Most patches for Vintage Electric Piano have a Bell and Drive control. Bell controls allow you to make ringing sounds more prominent while the Drive control adds a warmer distortion sound to your track.

Acoustic Piano patches offer Low and High controls that raise or lower either the bottom or top sounds. These controls also offer a deeper warmth and body to the sounds when used right.

5. Synthesizer

Synthesizers are used because they can emulate or mimic a mix of sounds. This means that the synthesizer is a unique instrument and creating a base control system is difficult.

Patches for synthesizers vary greatly and are based on an advanced synthesizer engine or they use a basic subtractive structure that includes the following elements; Oscillators, Filters, and Envelope Controls. Each of these subsets controls different sound creations.

GarageBand's Synthesizer control panel is composed of the Envelope controls that have filtering and oscillation controls like Attack, Decay, and Release.

Synthesizers also have effects such as Chorus, Flanger, and Delay.

6. Vintage B3 Organ

The Vintage B3 Organ has slider controls that allow you to adjust the volume and settings of a variety of harmonic sounds. The controls use drawbars to do so. To adjust the volume of any of the harmonic sounds drag the drawbar down or up. The drawbar is different from a normal slider as dragging it up will soften the harmonics and dragging it down will raise the volume of the harmonic.

The Vintage Organ includes effects like Reverb, Organ Verb, and Distortion. Distortion creates a tube amplifier sound and can create a warmer sound. Organ Reverb creates a reverb that mimics spring reverbs.

7. Vintage Clav

All of the Clav patches in GarageBand mimic the control switches that are found on the D6 clarinet. These switches are Brilliant, Treble, Medium, and soft. Each has a control function used to change the bass sound.

There are various effects that can be found alongside the Clav controls. Effects like Phaser Wah and Flanger help you fine-tune the clarinet sound that works best for your project.

8. Vintage Mellotron

The Mellotron recreates old-school keyboard sounds from the 1960s and 1970s. It is commonly known to be the predecessor to any modern sample-playback instruments.

Commonly the Mellotron has effects included like Amp, Delay, and Phaser. By adding sweeping sounds, echo effects, and providing a warmer tone and distortion these effects will help you create the classic 70's rock track you've always dreamed of.

To open any of the smart control panels make a track selection. The smart control panel will be shown in the bottom right corner of the main window. Here you can make all the adjustments you need to make to your segments. No matter which track selection you made, the basic layout of the smart control screen is the same. All tonal controls can be found on the left-hand side of the panel, while the effect changes will be located on the right-hand side.

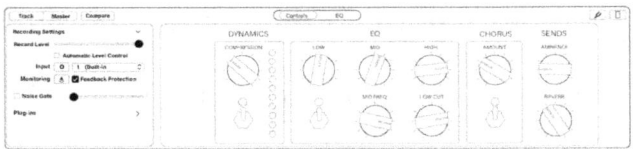

Amps and Pedals

In this section, I will briefly cover the amps and pedals that GarageBand offers you. GarageBand offers a variety of amps and pedals that mimic popular electric or bass guitar sounds. Once you make your track selection you will be able to select patches that include amps or a stompbox. You can use these sounds alongside vocals, drum tracks, and even electric pianos. GarageBand also offers you access to the Amp Designer, Amp Base Designer, and Pedalboard plug-in panels. These are located alongside the smart control panels. Both the Amp Designer and the Bass Amp Designer allow you to change the amp settings and make adjustments to the amp controls. You also have the capacity to choose a variety of cabinets and microphones to customize the sound even more. The Pedalboard allows you to reorder any effects stemming from the pedals and customize any pedal routing.

Amp Designer

Amp Designer mimics and copies the sound that popular guitar amps make. Each Amp in GarageBand includes the amp, the cabinet it needs to work properly, and an EQ that helps with sound recreation. GarageBand offers a menu with various amp models but also allows you to create your own amp. You will need to add in an Amp, Cabinet, and Microphone when creating your own amp model for usage. You will have access to 7 different microphone selections when creating your own amp.

The basic amp plug-in is constructed of the following items:

- Model Settings: This includes the model pop-up window where you select your entire amp model.
- Amp Settings: The outer knob settings area, the settings that can be changed here are the input gain, presence, and master level.
- Effects Settings: The center knob selection where integrated effects are controlled.
- Microphone Settings: The right position of the interface hosts these settings, allowing you to set the type of microphone as well as the position.
- Output Slider: Located in the lower-right corner, the output slider controls the level control for the output.

Please note that the Output Slider is different from the Master Control. The Master Control changes both the output and the sound design.

To open the amp designer simply select the track that makes use of the electric guitar patch and clicks on the dial icon and then the Amp Designer icon, which looks like a little briefcase.

Amp Designer has different Amp models to choose from. These models include an amp, cabinet, an EQ, and a microphone that has a basic setup. GarageBand offers the following Amp Models:

1. Tweed Combos

Creates a sound similar to 1950s and early 1960s rock, blues, and country.

2. Classic American Combos

Creates a sound similar to middle 1960s vintage R&B, jazz, or country.

3. British Stacks

Creates a sound largely defined by rock, and based on 50-watt amplifiers with 4x12 cabinets.

4. British Combos

Creates a sound similar to 1960s British pop and British Rock.

5. British Alternatives

Creates a sound similar to 1990s British Pop sounds that can be played at high levels without note distortion.

6. Metal Stacks

Creates a sound similar to modern hard rock and metal with a heavy distortion sound for long sustained guitar tones.

7. Additional Combos

You can use multiple amps together to create a versatile sound and music style.

To create your own amp model for usage you will need to follow a few steps. It is rather easy to do so and you can save your amp in the library once you have finished creating it. You will need to use the Cabinets, Amps, and Mic pop-up menus to create your own. You can change the window size by pressing the triangle icon in the lower right-hand corner. Amp creation will be covered very briefly in the following section.

You will need to follow these steps to create your own amp. You can do so by clicking on Custom EQ in the left knob section on the current amp shown:

1. Choose an Amplifier

Any of the above-mentioned amplifiers can be chosen.

2. Choose a Cabinet

Any of the following Cabinets can be chosen.

- Combos or Stacks
- Old or New Speakers
- Large or Small Speakers
- Single or Multiple Speakers

3. Choose an EQ type

Click on EQ and rotate the Bass, Mid, and Treble knobs to create your EQ.

4. Choose a Microphone and its placement

There are three types of Microphone models to choose from:

- Condenser models
- Dynamic Models
- Ribbon 121 Model

Move the Microphone by hovering your mouse below the cabinet until a grid appears. Simply drag the white dot (simulating the microphone) to the position you want it to be placed.

To use the amp, simply adjust the different knobs to your liking. Finding the sound you want and getting comfortable with the controls and panels will help you find the sound you want. Don't be afraid to play around with plug-ins and change up the settings to find out how each knob will change the sounds you're creating.

Amp Bass Designer

The Bass Amp Designer is similar to the Amp Designer, the difference being that the bass amp has a direct box. The direct box creates a sound similar to directly plugging your instrument into a mixing board. You can opt to choose the Base Amp Designer with the direct box, or use them separately.

The Bass Amp Designer is constructed in exactly the same way that the Amp Designer is. It consists of Model settings, Amp settings, Effects settings, Microphone settings, and an Output slider. You can configure these settings how you wish them to be, similar to the settings for the Amp Designer.

To access the Base Amp Designer, simply click on the base track, then click the dial icon and the Amp Designer icon that looks like a small suitcase. The Base Amp, like the Amp, has different amp models that can be part of your selection. However, there are only 4 types of Bass Amps to choose from:

1. Classic Amp: Mimics a six-tube bass amp that was first introduced in the 1960s and can be used for a large range of sounds.
2. Flip Top Amp: Mimics a 300-watt tube that was first introduced in 1969, and is best used for fuller tones.
3. Modern Amp: Mimics a 12-tube, 360-watt tube that was introduced in 1989, and is best suited for high articulation sounds.
4. Direct Box: Mimics a direct box to route the bass output to a mixing console panel.

Creating a Bass Amp is much easier than creating a normal amp. You can also create a custom Base Amp by following the following two steps.

1. Choose a bass amp

Choose any of the above-mentioned amp selections.

2. Choose a base cabinet

There are 8 different cabinet selections to choose from:

- Modern Cabinet 15"
- Modern Cabinet 10"
- Modern Cabinet 6"
- Classic Cabinet 8x10"

- Flip Top Cabinet 1x15"
- Modern 3 Way Cabinet
- Direct (PowerAmp Out)
- Direct (PreAmp Out)

The cabinet selection you make has a big impact on the overall sound of your bass amp. Feel free to play around with the settings and create sections of both the Amp Designer and the Bass Amp Designer.

Pedalboard Plug-Ins

Pedalboards in GarageBand simulate the stompbox pedal effects found with electric and bass guitars. Signals run from left to right in the area and allow you to change the flow. By adding busses, splitters, and mixer units you can create your own pedal combinations for an amazingly unique sound.

The pedalboard layout is constructed as follows:

- Pedal Browser: The right-hand panel is where you can view all the pedals that come pre-installed in GarageBand.
- Pedal Area: The main pedalboard is where you can add, remove and reorder any and all the pedal boxes found in your project.
- Router: The upper part of the pedalboard and used to control any signal flow as well as the two effect busses.

To open the pedalboard hit the Smart Controls icon that looks like a dial and then hit the Pedal icon that looks like a pedal. Note that you can only access the pedalboard if it has been plugged into the track selection you have made.

GarageBand offers a variety of pedalboards that each create their own sounds and effects. The following pedalboards are included in GarageBand, they can also be changed and edited:

1. Distortion Pedals: Creates louder signals and includes fuzz and overdrive effects.
2. Pitch Pedals: Great for pitch shifting and includes octave and wham effects.
3. Modulation Pedals: Creates modulation and adds swirling sounds, also includes chorus, phaser, and flanger effects.
4. Delay Pedals: Great for playback to make ambiance changes, includes echo effects.
5. Filter Pedals: Used for frequency emphasis and includes wah and EQ effects.
6. Dynamic Pedals: Used for volume control and includes noise gate and compression effects.
7. Utility Pedals: Great for additional routing without affecting the sound, includes mixer and splitter effects.

To reorder the pedals in your pedalboard, simply drag them around to reorder them. You can also remove and replace pedals. To remove and replace the pedals drag the stompbox so it is located directly over the pedal you want to replace and double click the pedal you want to replace it with. To simply remove a pedal click it and hit the delete button, or drag it from the pedalboard to the pedal browser.

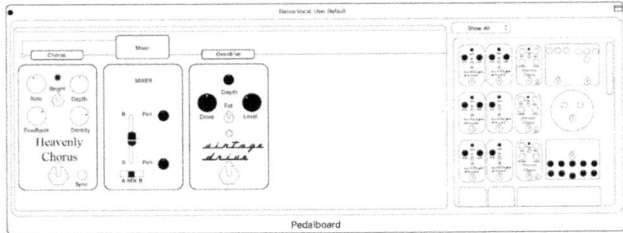

Mini Tutorial 6

This tutorial section will use the content covered in this chapter as well as the chapters before this one:

1. Using shortcuts open the 'I've gotten this far' file.
2. Rename the file to anything of your choosing.
3. Save your file.
4. Go through the mixing steps and use them to make changes to your file.
5. Consider adding Amps and a Pedal to your project if it is necessary.
6. Save your file.

This chapter's content relates closely to the content found in Chapter 1.

ENHANCEMENTS

In this section I will be covering the final enhancements you can make to your track before saving it and exporting it as a final project. While there will always be something you could go back and change, there comes a time when you make your final adjustments and export your file. This Chapter will cover the information you will need to double-check and can alter right before finishing the project to ensure that it sounds amazing and does not lose quality.

I will cover 5 of the most important tracks to look out for and how their adjustments will change your entire project. All of these tracks can be used in the tracks area, audio, piano roll, and score editor sections. When using them they will be shown directly underneath the ruler at the top of the screen.

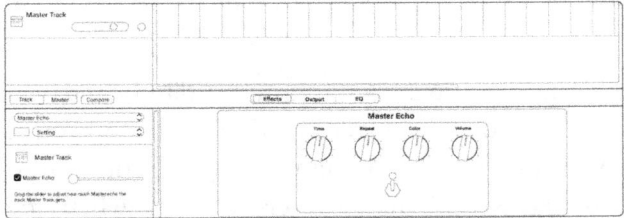

Master Track

The master track has control over the entire project and mainly focuses on controlling the playback of the track, including items such as fade-ins and fade-outs. The master track generally appears beneath the final track in a project. It controls the final playback volume for the entire project.

It is best to adjust the playback volume in the master track to a level that is high enough to cancel out background noise, but low enough to not cause clipping. Adding a fade-out to the project will add 4 volume points to the master track.

To open the master track, select the track, and select Show Master Track. To add an automatic fade-out select the track, click on Mix, and select Fade Out. You can alter the length of the fade-out or manually add a fade-out. I suggest adding an automatic fade-out to make things easier on you. To alter the length, have the master track open, choose volume in the menu selection in the master track header, and adjust the volume curve for the end of the track.

Arrangement Track

The arrangement track has all the arrangement markers and is used to control all the arrangements in the project. Arrangement markers are used to create different sections like the intro, chorus, verses, and outro. Arrangement markers start at the beginning of a project and will continue until the final seconds of sounds. Every arrangement marker that is added, starts where the previous marker ended. The best way to use these markers is to split your project into sections, allowing you to move it around without destructive editing and changes.

Arrangement markers can be adjusted, edited, and renamed to reflect the section that it currently is attached to. You can move arrangement markers and edit them as much as you want or need to create the perfect sound. It is especially useful when you wish to copy content like the chorus without having to rewrite the section entirely. You can view the arrangement track by selecting a track and choosing the Show Arrangement Track menu item. It will show up directly underneath the ruler in the main window.

Tempo Track

The tempo track hosts all the tempo and tempo changes for the project. The tempo by default is generally 120 bpm (beats per minute) and can be set when you choose a track or using the LCD later on. Tempo tracks can be edited and changed in the Tempo track window. To view the tempo track select Track and choose the Show Tempo Track menu item. This will show the tempo track below the ruler.

You can adjust the tempo in the tempo track by dragging the start and end points up or down. By dragging them down, you will lower the tempo and by dragging them up you will push the tempo higher. A higher tempo means a more fast-paced song, and a lower tempo means a slower-paced song.

When viewing the tempo track you can adjust the tempo by using tempo points. You simply double-click and GarageBand will add a tempo point to the selection. Here you can drag it higher or lower. You can also add a numeric value manually by holding down Control + Option + Command and clicking in the spot where you want the numeric value to be added. Enter the number and hit the enter key.

You can also add tempo curves to your tempo track. You can do this by selecting the two points where you want the curve to be, clicking and dragging the mouse pointer above or below the second point. This will create a curve that you can drag horizontally and/or vertically to change the shape and tempo of the section. You also have the option to move, copy or delete the tempo points. This you can do by using the shortcut keys.

Transposition Track

The transposition track includes all transposition information and history that has been made within the project. To transpose a track or section means changing the pitch of that section. You can do this for every track and selection to keep the pitch on your project perfect.

Transposition has an effect on software instrument tracks, as well as Apple Loops and MIDI regions. Audio regions are not affected unless you have ticked the Follow Tempo and Pitch tick boxes in the corresponding editor panels.

To show the transposition track for editing, select Track and choose the Show Transposition Track menu item. The track will show directly underneath the ruler. You can add, remove and change the value of transposition points in this panel. Command-clicking will add a transposition point, dragging it up or down will change its value, and using the delete button after selecting the point will remove it.

Movie Track

The movie track hosts the film and the frames of the film that have been attached to the track and are synced to the ruler for scoring. The content regarding this was covered in Chapter 4 when I covered Media and how to add films to your tracks.

Mini Tutorial 7

This tutorial will cover all the content found in Chapter 7 and all the chapters that came before it:

1. Open your file.
2. Add any track, regions, amps, pedals, or plug-ins that you would like to add.
3. Go through your project by using the above-mentioned track selection by opening each track and doing some minor editing. (Hint: End with the Master Track)
4. Save your file.
5. Export it to your chosen format.

The content covered in this section closely relates to the content covered in Chapter 5.

NEW SKILLS

In this section, I will briefly cover the Lesson section that GarageBand offers you. They offer an amazing program that can be structured to suit your needs. The Apple Lessons section offers you classes in guitar and piano playing. The Lessons section also allows you to download extra lessons.

These lessons are called Artist lessons and are taught by the artists who created the tracks themselves. These lessons can be downloaded before or after you have started playing the instrument. It all depends on your skills.

The lesson sections are split into 4 different sections. Some chapters will have subchapters that include simple and advanced techniques and skills to learn:

1. Learn

The 'Learn' chapters are the sections where the instructor will teach you and guide you on how to play the selected track, also teaching you any techniques you may need.

2. Play

The 'Play' chapters will be the section where you can play the song without having an instructor present, or you can practice individual sections.

3. Practice

The 'Practice' chapters will offer you exercises specific to the track, where you can learn specific riffs and techniques to help you learn to play the song.

4. Story

Artist lesson chapters include a backstory about the song and will also have some guidance from the artist to help you learn techniques specific to the song.

Lessons

In the main window once you have opened GarageBand you will see the Lessons sections in the left-hand pane. Once you have clicked on this section a new window will open. You can either select Guitar Lessons, Piano Lessons, and Artist Lessons. When selecting Guitar or Piano lessons, the first lesson will show up. All the lessons that have been downloaded will show up underneath this section.

If you double-click the lesson, it will open into a full view screen. An animated fretboard (for guitar lessons) or keyboard (for piano lessons) will be shown at the bottom of the screen, along with some video instructions. Chords, music notation, and notes will be playing in the center of the screen. The screen also has various controls for playing, volume, and pausing the lessons.

The triangular icon in the control bar will play the lessons for you, you can also use the spacebar to play and pause the lessons. There is a panel to the left of the lesson screen where subchapters can be selected. When clicking on the 'Learn' section, you will get the key instructions for the lesson. Clicking on the 'Play' section you will be able to practice the song alongside the song playing in the video. If a practice lesson is included in a chapter, you can click on it to learn the chapter-specific techniques. Chapters that have subchapters, can be clicked to view the instructions for those sections. In Artist chapters, you can click on the 'Story' tab to view more information about the song or the artist.

To go back or move forward in a lesson, simply move the playhead by dragging it left or right. The lessons are split into subsections, to move to a different section, simply click on the section in the navigation row. The playhead will move to the beginning of the section. If you feel the need to repeat a section, simply click on the section, or drag the playhead back to the section which you want to repeat.

By selecting the 'Play' section, you will be able to play the lessons without instructions. You can play, pause and rewind as much as you need to during these sections. Once you have finished your lesson, you can close the window by pressing the small 'X' in the top left-hand corner.

Instruments

If you want to play guitar lessons, you will need to connect a guitar to your computer before being able to play the lessons. The best method to do so is to set up your guitar as an input source in the setup window. The setup window is located in the top right-hand corner. Once the settings window opens, use the drop-down menu to select your device.

Certain inputs are far better for certain instruments. You can select one of the following inputs:

- Internal Microphone should be used when you're using an acoustic guitar and using the MacBook's built-in microphone.
- Line-In should be used when your guitar is connected to the audio input port found on your computer.
- If you have opted to connect your guitar to a USB device, select this device in the drop-down menu.
- If you have opted to connect your guitar to an Audio Interface device, then select this Audio Interface device.

If you are using an external microphone make sure that you toggle off the monitoring to stop any feedback from happening. Also, check your volume slider to avoid any clipping and distortion. Once this has been done you can just close the setup menu and return to your lesson.

You also have the capacity to change the view of the window and to customize how notation is shown, if it is in fact shown or if there is an animated instrument in the lesson window. By clicking on the note icon you will be able to select one of the notation view options in the drop-down. Selecting the Full page notation option will allow you to see the music notation in a full pageview.

Review and Progress

GarageBand also offers you the capacity to track your progress, both in real-time and after your lesson has finished. After a lesson has finished you can view your scores in the Progress and High Scores windows. In both windows, you will be able to view your overall scores, and more detailed information can be viewed on individual performances.

In the main lessons window, in the bottom right-hand corner you can access the History window. A graph will show the progress you have made, by clicking on a point in the graph, it will bring up the progress for the individual lesson.

To view your highest score simply select the menu in the top right-hand corner. The scores will appear in bars with the percentage score to the left of the bar. You will be able to review each progress bar, just by clicking on the review button below the bar.

You will also be able to change the mix of the sound in your lessons. This is especially helpful if you want to turn down the instructor's voice or turn up your own instrument to hear better. To be able to change those settings, click on the Mixer icon in the top right corner. Change the volume as needed and continue with your lesson as follows.

You will always be able to go back and redo chapters or sections. Don't be afraid to pause lessons and focus on getting comfortable before moving on to the next section or lesson.

PART II

INTRODUCTION

Welcome to the second part of GarageBand Basics. Congratulations on getting all the basics down in the first part! It definitely was hard work! You can be proud of yourself, and pat yourself on the back for getting that information down. I hope that the information in this second part can help you further your musical production career even more. In this part I will cover some familiar concepts, but I will deep dive into them, allowing you a better insight into the arrangements, the controls and even using GarageBand on your iPhone.

This part will start off by quickly covering the basics again before moving on to the in-depth knowledge you will need to create perfect arrangements, using the best controls for your project and how to record any sounds or instrumentals to create an even better sound. The final chapter will cover GarageBand mobile. This chapter will cover all you need to know about using the mobile version for on the go editing and recording.

As with the first part, I will be including tutorials. Unlike the first part, there will not be mini-tutorials, instead I will be creating tutorials for each section that I cover, making sure that you have each section down before moving ahead will greatly improve and help you understand the reasoning behind each section.

I hope that this part, like the first, brings you a wealth of knowledge and helps you build up your confidence to release that new hit single or submit that self-made film to the art school of your dreams!

I wish you good luck on your journey! May you be as happy and passionate as the day when you first fell in love with music!

PART 1 RECAP

In chapter one I will quickly cover all the basics that were covered in the previous part. This chapter is a look back for all the information you might need to refer back to later in the guide. This chapter will cover the basic concepts that were covered as well as quickly gloss over the content that will be covered in this part.

The first part of the GarageBand guide offered you information on how to navigate the software as well as how to get around most of the areas without having to know the inner workings. While the first part was wonderful in guiding beginners, knowing how the mechanics work in GarageBand will allow you to better create and visualize your project. Only the pure basic usage will be covered in this chapter before moving on to the more in-depth details that will be covered in the next chapters.

Basics

This small section will cover all the basic usage inside the software as well as the main shortcuts that you will be using. As mentioned in the first part, GarageBand offers its users a variety of pre-recorded sounds and instruments to simply select and plug into any project no matter how big or small. As mentioned in the first part there are some steps that need to be done before you can use the software.

When you first open GarageBand after downloading the software, you will be prompted to download the loops library. You can opt not to download any additional loops and track but this might leave your sound library significantly smaller than what you would like. The download is rather large and I would suggest downloading it over a strong connected network. Once you have downloaded the loops library you can continue, however if you find yourself stuck without certain sounds or wanting something more diverse you can opt to download the entire GarageBand library for Mac. These sections will of course be a lot smaller if you use them on your iPad or iPhone, and the prompt might look different. The choice will stay yours. I will discuss the download process and first views of mobile GarageBand in Chapter 5.

Now that you have downloaded all the necessary files you can move on to scoping out the look and feel of the software. If you have worked with GarageBand before, or seen some images of what the software looks like you will be okay to move on to the next section of this guide. However, if you haven't used the software before or its been a while since you have used it, then the following section is for you. Once you have downloaded the files you will be using, Garage-Band will open with a menu pane to the left-hand of the screen. This menu pane allows you to navigate the beginning stages of any project, or lets you take some lessons if that is your goal.

The menu pane includes the *New Project, Recent, Learn to Play, Lesson Store* and *Project Templates* menu items. Any one of these menu items will guide you through your choice. The *New Project* menu will open a dialogue box where you will select which track to start your project off with. The *Recent* menu allows you to open any recent projects you may have worked on. The *Learn to Play* and *Lesson Store* go hand-in-hand. The former allows you to learn how to play either guitar or piano while the latter is the store where GarageBand allows you to buy specific lessons for the instrument you would like to play. The last menu item, *Project Templates* allows you to use predetermined tracks for your projects that have been grouped together to give you the best possible ensemble.

The best way to start a new project would be using the *Choose a Project* window to the right of the screen. This will allow the user to select the Tempo, Key Signatures, Time Signatures, the Input device (mic or instrument) and the Output Device. Most of these settings will be preloaded as the system default and are often the best used ones that you can find. It is important to remember that these settings are not finite and can be changed once you have started your project, they do however have an impact on the project if they are changed.

The next most important thing to remember would be the shortcut keys. These will save you a lot of time, especially since a lot of the menu options these represent are only accessible via the right-click menu, or the shortcuts. You will find all the system shortcuts in the shortcut section that has been included after the conclusion of the book. There are a few shortcut keys that are vital to ease-of-use in the beginning of the project, no matter your skill level. I use these all the time whenever I am working on a project, and they become muscle memory after a while.

Shortcut	Action
R	Starts the recording process
Space Bar	Play or Pause the current recorded section
Command + S	Saves the current project
Command + Z	Undo the action just completed
Command + X	Cuts any selection or text
Command + C	Copies any selection or text
Command + V	Pastes any selection or text that has been cut or copied
Command + A	Selects all the tracks or info in a cell

These shortcuts are a lifesaver and more often than not I am using these to start my project off. There are however other shortcuts that are used in the various editing and track selection screens but you can find those in the *Editor Shortcut Key* and *Global Track Shortcut Key* sections after the conclusion.

Projects and Recording

The projects and recording section is rather easy to explain as this is the section that covers how to set up a new project as well as how to go about recording your own sounds. This section goes hand in hand with the tracks and plug-ins section. Track selection will help you create the project you've wanted from the beginning. Once you have filled in the basic information on the *Choose a Project* window, the program will prompt you about the starting track. The starting track will be one of three tracks. These tracks are color coded in GarageBand, allowing you to easily identify all the information as well as plug-ins and patches related to those specific track types.

Track types can be identified as follows; *Audio Tracks, Drummer Tracks or Software Instrument Tracks.* Audio tracks are generally green in color, while Drummer Tracks are yellow and Software Tracks are green. The color differentiation allows you to add the correct loops, plug-ins and other color related media to the correct tracks, allowing you to keep track of any section of information effortlessly. Each track type has its own recording settings and can be changed within the track section once you have selected the track in the project window.

When you are starting a new project, simply select the basis of your project, whether that is a Drummer, Audio or Software track is up to you. Once you have selected your track type the software will automatically load you into the project screen. On the left-hand of the screen you will find the library selection, where you can change your track selection, whether you want something happy and cheerful or something more alternative like punk and rock, there will be an option for you. GarageBand also includes an audio track made for narration, allowing you to record your podcast, book narration or even film narration. Most of the upper right-hand screen will be taken up by the track area. The small settings block to the left of the track, and to the right of the library will allow you to change and manipulate the recording settings for these tracks. Each track type has its own settings and impacts the track in a different way. Don't be afraid to play around with the settings, I find it easiest to learn how something works or changes by doing different things with it. This allows artistic freedom to create truly unique sounds.

GarageBand also offers its users the capabilities to change both the metronome as well as the tuner. The tuner is an amazing tool that allows you to tune your guitar to play any notes on the right tune. The Metronome allows you to play in time with any pre-recorded or other tracks that you have loaded into the project, allowing you to be on time with the sound. What also makes the metronome an amazing functionality is the fact that you can use it along with a count-in before recording your own tracks, making it that much easier to record your own tracks without having to fumble between hitting the record button and playing your track. Along with the metronome you will have to use the count-in, this function counts a few seconds before automatically starting the recording process. The metronome will keep track of the pace of the song with the ticking sound that it makes while the count-in does the work for you, allowing you to focus on the creation of your next masterpiece.

It is important to note that you can adjust the settings for the metronome, these settings include the volume as well as the tone of the metronome, making it either louder and more high pitched or softer and deeper. These settings can be found in the preference menu, and metronome submenu.

When it comes to recording it is important to remember that each track type has different settings to change when recording. Settings include the *Record Level, Input and Monitoring*. Each of these settings change the output of the track and overall output of the track. These settings can also be left at default and once you move on to the editor section, can be changed globally to change the feel of the entire track, instead of just changing one section of the selected track. Using the built-in editors will allow you far more control over your tracks, and allow you to refine them to the smallest of sections. Editor changes allow you to change the pitch of a single note in a recorded section, or change the flow of one sound to the next without altering the entire track or sound. These settings go hand-in-hand with the arrangements as well as the controls section.

Media

The media section can be used to create both film and book narration. The media section of GarageBand allows you to create audio files for films, by re-recording the soundtrack for the film. This allows you to create new sound for the film and replace the original soundtrack. It is important to note that these changes have to follow a specific method to keep the sound and film in sync. These settings and steps were explained in the first part of Mac GarageBand.

The media section also includes most of the Apple Loops library. Apple Loops are snippets of pre-recorded sounds that can be arranged the way you wish to create the rhythm, style, tempo or base of a track. Apple offers a wide variety of loops that can be found in the library, and that are downloaded when you get the first initial prompt to download the all inclusive library of sounds. What makes loops so amazing is the fact that you do not have to use the pre-recorded GarageBand loops, you can either create your own loops, or find Apple verified loops to use as a plug-in, alongside your project.

Loops are a convenient way to create a unique snippet of sound that can be used over and over throughout a project, whether It is a vocal loop, a guitar riff or a drum beat. I cover how to create your own loops in part 1. Loops are also convenient because you can use the shortcuts to use them wherever you want in the track. Remember that *Audio Loops* can only be used alongside *Audio Tracks*, the same can be said for *Software Loops and Tracks*, as well as *Drummer Loop and Tracks*.

Loop customization is a perfect way to create a range of unique sounds within each track. GarageBand also offers you the capacity to ' favourite' your preferred loops, allowing you easy access to any loops that you might want to use throughout a project or in the future. GarageBand also allows you to create your own loops, allowing you to record a sound of your liking and looping them. These custom loops are automatically saved within the loops library, allowing you to use them in future projects or adding them to your favorites list.

When it comes to media imports for any type of files, GarageBand only allows 6 file types to be used within the software. These file types are: AIFF, CAF, WAV, ACC, Apple Lossless and MP3 files. These are the six main audio file types that can be used alongside your media that has been imported. You can also import MIDI files into GarageBand if you prefer to use those instead. Unfortunately you can only use one film file per GarageBand project, limiting your usage in GarageBand. I would suggest editing your film footage to the extent that you only need to add the sound and then importing the file into GarageBand, making it easier to work with, and also allowing you the capability to sync the sound much better. Once you have synced your film to the desired soundtrack you can export it using the *Export Audio to Movie* selection and you'll be able to show off to all your friends that you made a film.

Arrangements

Arrangements are the small changes and track selections that make up the entirety of a track or audio output. These arrangements allow you to create the unique sounds that we find in film, audiobooks and on all the videos we find on YouTube. Arrangements allow you to layer specific sounds, as well as lowering the volume for some selections while changing the pitch of others. Utilising the full extent of arrangements allows you to better create a comprehensive full sound instead of chunks of sound that do not blend well together.

GarageBand offers you editors for tracks to help you better navigate the changes you want to make to every loop or track to make the entire arrangement as perfect as you need. Arrangements are split into regions. Regions are the small sections of sounds that are used without the track. The three seconds of rain, four seconds of thunder, the 60 seconds of voice audio, are all different regions. These regions can be looped, cut, or paste and even edited using the editors for their specific track type. Each track editor has different settings that can be altered and even allows for note, timing and tempo changes in one region, or even a track type. This is covered in part 1 but will be covered in Chapter 3 of this part as well.

Alongside the Editors you also find the capacity to write music notation, or even just make notes. This is especially useful if you're in a rush or working on a project with multiple people. I love keeping notes on tracks, as well as the editing I'm doing to each arrangement or region, this allows me to go back and check the changes I made if I haven't worked on the project for a few days or if I haven't been able to create the sound I like but I finally do, I'll make a note on how I made a specific sound. This allows me to go back to that project and see how to create a certain sound.

My favorite part about GarageBand's music notation is the fact that you can export your music notation from GarageBand and print it out, allowing you to take it with you or even sending it to someone else to record it on an instrument in a studio or on the go, using the GarageBand mobile version.

Controls

Controls are amazing plug-ins that allow you to alter and change the sound of your tracks even more than any editors or tweaking can do. Controls, along with Smart Controls and automations allow you to create the best sounds within your project. GarageBand contains a large amount of pedals and amps that can be included in your project. Automations are also linked to mixing in GarageBand. Mixing can be split into four different steps that are to be completed before moving on to Automation. These four steps are as follows: Volume setting, pan position setting, effects and finally automation curves.

Setting the volume will allow you to set the volume of each individual region and track. Pan position setting allows you to set the panning position for each individual track, allowing you to set where each sound will be heard on headphones. Effects are used for different reasons, sometimes It is about changing the effect to sound like It is in a deep dimension where sounds echo, other times you would want it to sound more robotic. Automation Curves are the starting base for any automation that might be happening in your project. They help you change the smallest of details in your tracks, even a singular note in an entire track can be changed using the automation curves. Automation curves are also used to change and alter volume for certain regions or tracks as well as panning, echoing and reverberation.

Smart controls allow you to control the individual instruments that are within your project. GarageBand also offers you the amazing capability of importing your own, or other custom made controls for whatever sound you need. GarageBand has various different default controls that can be changed in the smart controls section; controls such as *Compression, Reverberation, Modulation and Noise Gating*. Each track has its own smart controls and while they may have the same name their changes alter each track differently. Like I mentioned before, the best way to get used to the controls that are found in GarageBand is to play around with them, change them up in the most unconventional way and go from there. You also have the option of looking up articles and blogs on how to better change smart controls to create the sound you're so desperately looking for. Most creators will include step-by-step instructions on how to change your smart controls to copy the sound they created.

Smart controls will be covered more in-depth in Chapter 4 of this part, and I hope to create a deeper understanding of how each change can alter the way you see and use these controls for your project.

Amps and Pedals are a wonderful way to control the *Drummer Track* in GarageBand. These settings and plug-ins allow you to create the perfect drumming sound without having to rent out a studio or pay for someone's time and energy to create your sound. GarageBand's *Drummer Tracks* are pre-recorded tracks from music professionals that allow you to create that authentic drumming sound that comes from watching a live show. Basic amp plug-ins have 5 default settings that can be adjusted and I suggest changing each of them and seeing how they influence the sound you are trying to create. GarageBand offers an Amp Designer that allows you to create up to seven different drum sounds within your project.

Amp Designer can also be used to alter the sounds of the guitar tracks you are using, to either create an acoustic sound or to create a bass sound. Amp Designer basically uses Pedalboards to change from acoustic to bass sounds, allowing you more musical freedom to create either a more country folk style track or a pop-rock album.

Enhancements are amazing small changes that can be made when you are close to finalising the entire project. These enhancements can be made to the master track, arrangement track, tempo track, transposition track and the movie track. It is important to remember to make your enhancements on the basic tracks, because any changes to the master track will change the entire project as the master track almost works as an override for any project changes.

All of these sections have been briefly covered in part 1 of GarageBand, but will be covered far more in-depth for this part. The tutorials for this part will also be far more intense and allow you to create an entire short film, or narration project for an audiobook. These have been planned out to allow you to create whatever project you need, as the first part allowed you to create a musical track. I hope that the tutorials will benefit each and everyone that chooses to use this guide to better their knowledge in both GarageBand and all the software capabilities in the software.

While most of the basics were covered in the first part of GarageBand, I want to use the tutorials in this part to take you step-by-step through the process I use when creating an entire track. I will include screenshots and graphs of all the items and settings I am using and allow you to follow step-by-step how I record and finish a single chapter of an audio book.

PROJECTS AND RECORDING

Project and how you go about recording any sound you need to create for whatever you are doing is one of the most fun things to do. While the initial set up and creation process might be tedious the results are always far more rewarding than one can imagine.

As mentioned in the first part you can use patches and plug-ins alongside your tracks, allowing you a wider variety of sounds and loops to play around with. You also use different recording techniques to get the sounds you need. plugins work hand in hand with the Pedalboard and Amp Designer and will be discussed in Chapter 4.

Patches and Tracks relate closely together and being aware of all the changes that can be made to either of these before even attempting to record might help you better understand the recording process.

Adding media to your file is easy and simple and while there is a lot of different media you can add, this chapter will not cover the basics of media, but rather the audio that goes along with the media selection you made.

Loop customization is also important when making new tracks. It is important to note that loop customization is a very powerful tool and can help you achieve the best possible sound no matter what, the only thing of course holding you back would be your knowledge on how to use the software.

Patches and Tracks

Patches are specific tones and feelings you want each track, or selection to make. These are found in the library, on the left hand side of the main screen. By selecting a patch, a selection of settings will be added to that track selection, and any recording made will have those settings attached to it.

If you wish to record an audio track with vocals, but you want the vocals to sound more experimental or alternative you would select the 'Experimental Patch' as well as the 'Monster Vocal'. This will allow you to create a sound that sounds like it is a monster singing or talking rather than your own voice. These patchers are a great way to diversify and change your tracks.

Recording

When you are recording it is important to know exactly how and which settings to change. These settings will have a great impact on how your track and project will turn out. While you can later adjust and edit these settings, it will be much easier to do so when you have set up your settings to be exactly the way you would need to have your track sound.

I suggest taking the time to record a small snippet of vocals or guitar riffs and then playing around with that section, and the settings related to it, before recording the final piece of the track. This will allow you the capacity to set up the correct settings and details beforehand and make the recording process a breeze.

While the settings are important it is also important to note that if you do not have the correct instruments or tools to do these recordings your sound might end up falling flat and you might struggle to record the sound you would like to.

Before you start recording, check that you have set up the tuner and the metronome as explained in part 1 of GarageBand. This will smooth out the process and not leave you stressed and anxious before the recording process has even begun. Once that has been done you can move on to selecting the patch you would like to use for your project. Now of course you will not always use the same patch, and odds are you might even have to tweak settings on the chosen patch so you can use it, but that is perfectly fine. Like I mentioned before, record a small snippet of yourself speaking into your microphone and change the controls for that track to your desired settings and then hit it off.

It is important to remember that if you are only working on the vocals for an audio track, that you can record in various different ways. I have found that knowing the shortcuts for recording is extremely handy, especially when I use GarageBand for audiobook recording. I will create a track, set up all the sounds, items and settings I need and hit the recording button when I'm ready. If I mess up a section I'll simply hit the recording shortcut again to stop the recording. I will take a breather, and then add a new track, that's exactly like the previous one, simply by duplicating it and then recording again. This allows me to have multiple takes of the same sections where I can pick the most favourable one and use it in my final track.

Loop Customization

Loop customization is extremely handy, especially in the case that I mentioned above. Having multiple recordings of the same section of text or audio allows me to customize that loop, or region the way I want, even if I have set up settings beforehand. This allows me to save a certain sound I made, or a certain riff I played, that was freestyle and I don't exactly know how to recreate it.

Tutorial 1

This tutorial will take you step-by-step through the first part of my project. From track selection to patch selection as well as setting changes and the first recording section to creation of loops.

Step 1:

The most important part of any project is planning what you want to do, while creating music and anything creative is a little more difficult to plan, I suggest sitting down and making notes on what you are working on.

For this project I am going to be narrating a chapter from my favorite fantasy novel series. Sitting down and reading the chapter beforehand allows me to make notes on the feeling I want to convey throughout the track.

Step 2:

Now that you've made notes and decided what you want to do, you can start up GarageBand. Be sure to download all the library sounds beforehand so you don't have to sit and wait for the downloads to finish when you are actually wanting to work on your project.

The next step is to select your project. You can let the default settings be used when it comes to tempo, key signatures, time signatures, and the input and output devices. The best possible selection when it comes to track selection to create your project would be to select an audio track. It allows you the most mobility when it comes to recording the sound you would like to create.

Once you have selected the audio track the normal view will open like always and you will be met with the library on the left hand side and the track view on the right. The bottom right screen will have the controls but you can ignore those. We will only be looking at the patch selection for now.

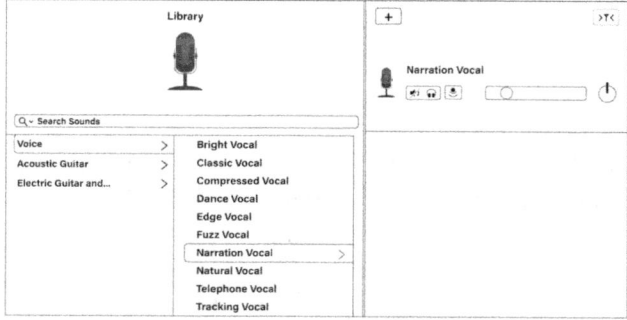

As seen from the image above, the only selection you will need to make is vocal, and then the appropriate vocal selection from the list of vocal patches. On the right side are the track controls. These can be left alone. I chose narration as I will be using this project as a way to narrate an audiobook.

Step 3:

Step three is the easiest, this is where you start working on the recording process. I would suggest recording a short section of lyrics or just normal voice narration, and then playing it back. This will allow you to gauge the mic volume as well as your speaking volume, and any other adjustment you want to make can be made during this section.

Final Notes:

Take your time during this process, you can easily get over eager to record and end up sounding rushed or breathless. I've found that the excitement often gets to me and I end up being flabbergasted and rushing through the narration, making the sound muddled and messy. Feel free to play around with the patch controls and volumes to make sure you're creating exactly what you set out to create.

ARRANGEMENTS

Arrangements are the easiest way to make sure that the track you are creating is flowing together nicely and easily. These arrangements make it easier to keep track of the different elements that can be found within your track.

Most of the track related information was covered in part 1, and is self-explanatory to a certain extent. The tracks, their regions and loops are all color coded to the same colors. Audio tracks, regions and loops are blue, software tracks, regions and loops are green and drummer tracks, regions and loops are yellow, allowing you to determine which sound is being made by which tracks.

In this section I want to focus on the editor for each track, and deep dive into the capabilities they have when it comes to each track. The editors in GarageBand are extremely powerful and allow you a lot of freedom when it comes to track manipulation.

Editors

There are three different editors, one for each track, allowing you various capabilities for each track type.

Audio Editor

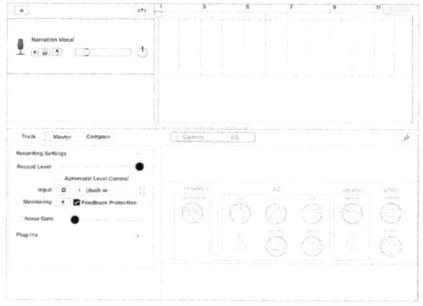

The audio editor is used to control any audio track, you can adjust these settings before recording anything or after you have recorded multiple sections.

The small section beneath the track that has the *Recording Settings* will allow you to change any settings when it comes to the audio editing and control changes.

These control changes will affect the entire track. Changing any of these settings will alter the sound on your audio track, and even on your master track. When using the audio editor I suggest zooming in on the track, you can do this simply by rolling the middle mouse button.

Zooming in on your track will allow you a better view of the track as well as the waveforms within the track. This will allow you to manipulate and change the specific regions and loops for each track. It will also allow you to edit and change the track on a smaller note level instead of altering a large section of the recorded sound and being unhappy about it.

Edits like pitch correction, and flex will allow you to correct the pitch on either the track or the selected notes. While the key limitation will allow you to correct notes and adjust them to a set note instead of correcting their pitch.

All these settings offer you in depth changes that alter the entire track, in the smallest and biggest way possible.

Roll Piano Editor

The Roll Piano Editor is used for all the software track editing that you will be doing. These tracks are mainly guitar, whether they be acoustic or base, and piano or keyboard sounds. Generally **MIDI** tracks are also included in this selection. The **RPE** (Roll Piano Editor) also allows you the capacity to edit the **MIDI** regions and to transpose them to your liking. Like the Audio track editor it offers you the capacity to edit any of the notes or regions, at a miniscule level, when zooming in.

The RPE also allows you access to the Score Editor where you can make changes to the score. These changes will affect your sound and track, but are also then changes in the Music Notation section, meaning you will not have to change the music notation before exporting the project.

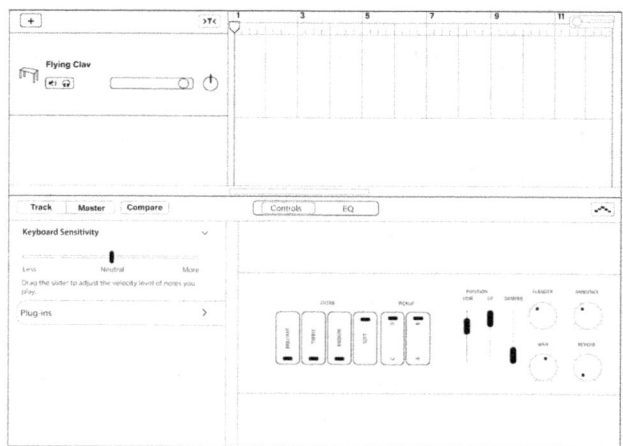

The RPE looks similar to the audio track editor, but the controls might change depending on the project or track selection you have made. The preview above shows a Flying Clav track, and the controls allow you to adjust the filtering on the treble, the brilliance, as well as the positioning, the dampening, the reverberation and the ambience. These controls greatly influence your project and track when adjusted so feel free to play around with the sound and listen to how these controls affect your sound.

Drummer Editor

The Drummer Editor looks vastly different from the PRE and the Audio Editor. Whilst it offers you the same editing and tools the difference in how it appears might seem daunting. The Editor is linked to the drummer controls and affects the entire drummer sound. GarageBand created the drummer tracks far more intricately than the other tracks as they are full pre-recorded tracks from drumming professionals. These tracks do not necessarily need changing but you can alter them in any way you want and need.

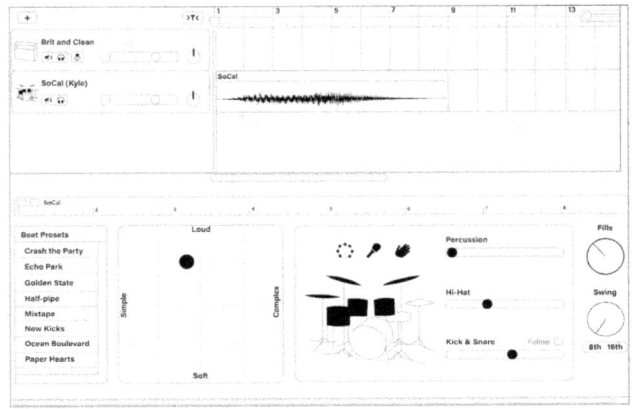

As you can see from the image above, the editor looks vastly different. Each setting in this control section for the drummer track will change the way the track plays back. These controls control the entire sound and will ultimately affect the entire track if your base sound is resting on this.

There are various beat presets that you can choose to use and even alter to create the sound you would like. For each beat preset selected you'll find that the settings to the right of the drummer image will change. It is important to note that when you select a drummer track it will automatically play along to the track you created, and the best way to create a drummer track that fits the way you want would be to create loops and regions within the trummer track.

The percussion, cymbals and kick and snare will automatically change when you select a different beat preset, but you can also drag them to where you would like them to be. You can also opt to lock the fills and swing on the two knobs that are located to the right of the percussion menu. Between the beat presets and the drum set image you will find a pan window. This window is dedicated to changing the panning of the selected beat preset and drumming track, allowing you an easier way to control how soft, simple, loud and complex a sound will appear when using those different settings.

It is important to note that these settings will only appear once, right after you have selected the drummer track, once you click on another track, and go back to the drummer track being selected the editor and controls will look similar to the Audio Editor and RPE.

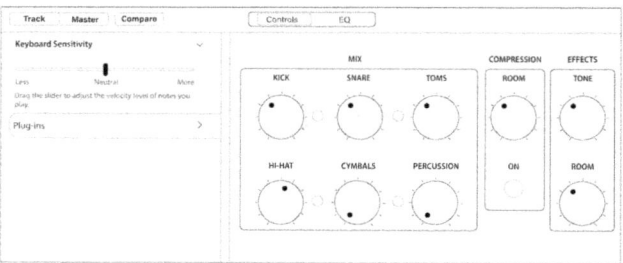

The above image is what the drummer track editor will change to, it might look different for every different type of track selection and drummer selection that you make. As with the first drummer editor, you can alter any of the controls to create a better or different sound than the one you are currently working with.

Music Notation

Music notation is important, whether It is just taking notes on the tracks so you can keep track of all the changes or if its having to share the track information with a creator so they can record the sound for you, it allows you to export your track information to a format that is more accessible than having to lug around your computer.

While you can view the Score, you also have a very handy notes tool available for you in the toolbar.

In the image above the first circled icon on the left would open the Score Editor for you, whilst the second circled icon on the right would open up the notes section. The Score Editor allows you to edit the notes on your selected track on a larger scale.

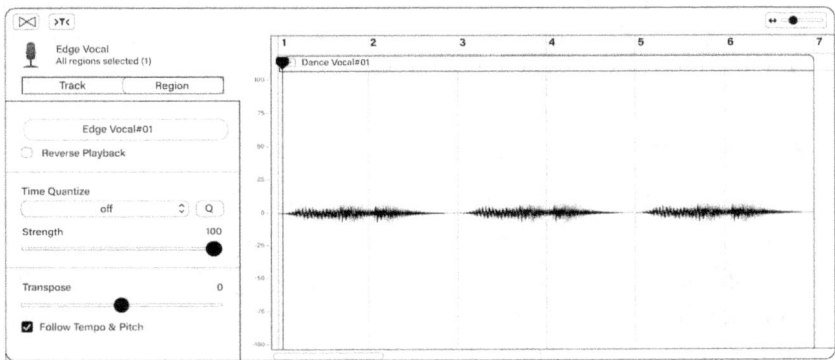

The score editor will look like the image above, recreating the waveforms of the vocals, as this instance is the narration track. Here you can change various settings surrounding the quantization and its strength as well as the transpositioning and the flow and tempo of the track. The score editor also allows you to change note pitch, length, velocity and quantization.

You can also do all the normal changes and edits that you would normally do, such as cutting and pasting loops or rearranging loops and regions as well as adding, moving and deleting any loops.

The Notation capability that is included in GarageBand allows you the easiest way to make notes surrounding your project. You can make notes with regards to tracks and even the entire project. They basically added the normal notepad into the software and made it conveniently attached to the project you are working on.

Tutorial 2

This part of the tutorial will focus on adding more tracks and starting to import and edit the tracks that I have been recording.

Step 1:

Since I know the manuscript I'm working with, I know that there are some dramatic scenes throughout the chapter. Since I want the audiobook I'm recording to be as dramatic and fun as possible I decided that I would add in some dramatic musical tracks. There is also a lot of singing throughout the book, and since I will not be selling the audiobook and it is a gift for a friend I thought I would buy the film's album and just use the singing from the album to create a more surreal feeling in the book.

To import tracks, you will have to make sure that the tracks are GarageBand friendly. In the first part as well as Chapter 1, I mentioned that you can only add 6 different types of track or media into GarageBand. I will thus be importing my tracks as MP3's since this is the easiest format to work with.

First I would need to add a clean audio track, I will open the folder where I have saved the MP3 files and then simply drag it into my software to allow it to sync to the track. I can now move the track forward and backward to wherever I want.

Step 2:

Once I have synced all of my tracks I always take the time to listen to the sounds together. This will allow me to gauge how loud a track is, and let me either turn it down in the track or the editor, allowing much more control over what is heard and what is not. I can also fade certain sounds in and out of the different tracks, letting the sounds flow far more harmoniously.

Once I have done this with the first track I can move on to doing this with the next track. Continue adding all the sounds you need to until you are happy with the base sounds and then go into the editors. I generally create my base track before going into editing as editing beforehand will only cause me to have to go back and constantly make changes. Once I have my base sound I can edit in the various editors to make sure I have the entire track down to what I want it to sound like.

Final Notes:

If you want to add your track to a film, this would be the perfect time to import your film to the software. Do this by using the menus at the top. Select the file option, scroll down to movie and click on the Open Movie selection. This will allow you to import your film into GarageBand. Importing a film will bring up a different view. The view will allow you to see the film frame by frame.

Seeing the film frame by frame will allow you to view the film in frame mode and make it easier to sync the track to the film. Oftentimes the track will not be as long as the film, to make sure that the track does not just stop in the middle I suggest adding a fade out and stretching it to be the same length as the film.

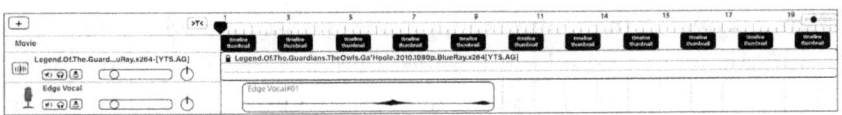

The film track will always be the first track, no matter what. This allows you to keep track of all the editing and changes you make while also seeing the changes in the controls window at the bottom or the score editor at the bottom. If you opt to make changes in the score editor, simply select the track and then click on the scissor icon that was mentioned in the music notation section. This will open the score editor at the bottom half of the screen and you will be able to make small changes.

Once you have finished editing you can save and take a break! You have deserved it. Be sure to always save continuously throughout a project.

Pro Tip: Click on the small file icon all the way to the left of the panel to close the library, making the tracks and controls area larger to make it easier to work with.

CONTROLS

Controls are one of the most amazing editing tools that GarageBand offers you as a creator. It really does define the sound and tracks a lot more than you think it would. This section will cover all the automations and controls that you can use within GarageBand to create even better sound and soundtracks within your project.

This section will also include the last part of the tutorial for the project. I hope that the explanations and step-by-step guidance has helped you create the perfect track. I can give as many tutorials and templates as I can but it really does all depend on how you want the track to sound or how you want it to playback.

Mixing is one of the controls that are the most used within GarageBand. Mixing can be split into four different steps, each doing something different. These steps allow you to make adjustments on a more detailed level. These steps are as follows; firstly, setting the volume, setting the volume of each track is important to keep one track from drowning out another, this allows each element of the final soundtrack to be appreciated and loved. The second step is setting the pan positions. Pan positions are important for any and all tracks. Generally, tracks are set within a stereo field. The stereo field has a middle, left and right section, these are important as setting your track to the middle will mean that the sound can be heard via headphones or earphone on both sides, setting a track or sound either to the left or the right will create that sound only on the side it was placed. Always double check your pan positions before moving on to the next step.

The third step is the effects, these are found in the controls panel below the tracks area. Each track has its own effects but they also play a big role in how the track is perceived by the ear. You of course don't have to use the effects if you do not want to but they add a lot of depth and character to tracks. General effects are things such as compression, reverberation and echo. There are many more effects that can be added to tracks and the best way to know what they do to each track is to play around and use them in your track and listen to the playback. I have said this a million times and I will say it again, the easiest way to figure out the controls and settings in GarageBand is to play with them within your project. Remember to save before each change so you can revert back to the previous track without having to redo your work. You can also choose to create a duplicate of your track in the same file or create a duplicate project to test all these changes. This way, if you end up losing something or changing something too much and you cannot roll back to the previous version you can always just delete the file. Always back-up all your files, no matter what you are working on. Save it to the cloud, or to a separate USB drive. You never know when your computer might crash and then you might lose the project you have been working months on.

The final step is Automation Curves (AC). ACs are extremely useful and I love using them, especially when I am working on an audio-book project. They allow me the freedom to change the tracks and edits on a note level. The AC can be added simply by hitting the A key on the keyboard. A yellow line will be created and every time you click inside the track a dot will appear. These dots are the Automation Curve icons. These can be dragged and moved around inside the track.

Automations

Automations are mainly used and adjusted using the Automation Curves. They are an easy way to create drops and heights in your tracks.

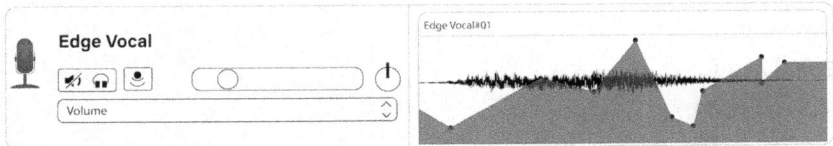

As you can see from the above reference there are 11 different AC's. These can be moved around simply by clicking and holding my left-mouse button and moving them. If you are done using the automation curves you can simply hit the A-key again and it will return to a normal track. The drop-down menu on the left hand side with the Volume selection will show what automation you are currently adjusting.

In the image above we added volume automation curves, to be able to add automation curves for the rest of the items on the track you will need to use the drop-down menu to select the item and then add your automation curve points.

This specific menu offers the following automations that can be added:

- Volume
- Pan
- Echo
- Reverb

The menu also offers you the capacity to add the following Smart Control Automation Curves:

- Compression
- Low
- Mid
- High
- Amount
- Ambience
- De-Esser
- Mid Freq
- Low Cut
- Reverb
- Bypass Channel EQ
- Bypass Tape Delay

You can also add Automation Curves for Plug-Ins in the following ways:

- De-Esser
- Channel EQ
- Compressor
- Exciter
- Tape Delay
- Channel EQ 2

It is important to note that each of the above mentioned plug-in automation curves each have their own sub-menus that host a myriad of options to choose from and each of these will have their own automation curve adjustments and changes when applied to each track.

All of the above information is only applicable to audio tracks, and software tracks and drummer tracks have their own automation curve drop down menus.

Software tracks have the following drop down menus.

Automation Curves:

- Volume
- Pan
- Echo
- Reverb

Smart Controls:

- Flanger
- Ambience
- Wah
- Reverb
- Low
- Up
- Damper

- Brilliant
- Treble
- Medium
- Soft

Plug-Ins:

- Channel EQ
- Compressor

The Drummer Automation Curves have the following menus.

Automation Curves:

- Volume
- Pan
- Echo
- Reverb

Smart Controls:

- Kick
- Snare
- Toms
- Amount
- Tone
- Hi-Hat
- Cymbals
- Percussion
- Room
- Kick on
- Snare On
- Toms On
- Hi-Hat On
- Cymbals On
- Percussion On
- On

Plug-Ins:

- Channel EQ
- Compressor

It is important to remember that even between the tracks there will be a difference in automation curves, smart controls and the plug-ins. I suggest double checking the drop down menus before making your changes, and applying them to all the different levels you want them added to.

Each automation curve that will be added will be a different color, allowing you to identify the different curves that have been added to the track. The below image shows the Volume (yellow), pan (green), reverb (blue) and echo (purple) automations that have been added to the vocal track.

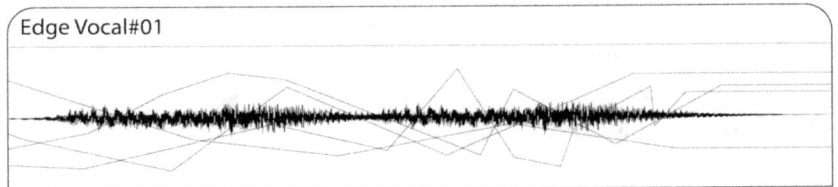

Smart Controls

Smart Controls are used to change and alter the way each individual instrument sounds within the track. They go hand-in-hand with the amps and pedals that GarageBand has to offer.

GarageBand offers 7 main effects or controls that can be adjusted, each affecting the track selection differently.

These main effects are considered the following:

- Compressor
- Delay
- Distortion

- EQ
- Modulation
- Noise Gate
- Reverb

These have been covered in the first part of GarageBand, and are pretty self explanatory. There are however 8 other controls that can be adjusted to your liking, each with its own adjustments and changes. These controls have also been covered in the first part so I will only briefly mention them.

Those 8 controls are:

- Guitar and Bass
- Drum Kits
- Orchestral and Mallet
- Vintage Electric Piano and Piano
- Synthesizer
- Vintage B3 Organ
- Vintage Clav
- Vintage Mellotron

Amps and Pedals

Amps and Pedals allow you to control and change the way that any track sounds. GarageBand offers a variety of different amps and pedals that can be used in your projects but if you want to create your own or import custom pedalboards you also have the capabilities to do so.

In this section I will cover the basics surrounding amps and the pedalboard and go over the designated pedalboard that can be found within GarageBand.

Amps

Amps are generally used with the string instruments and allow you some better tuning and editing to those selected tracks. GarageBand offers 7 different amp selections that all allow you to change the following settings:

- Model Settings
- Amp Settings
- Effects Settings
- Microphone Settings
- Output Sliders

The 7 different amps have been covered in the first part and are as follows:

- Tweed Combos
- Classic American Combos
- British Stacks
- British Combos
- British Alternatives
- Metal Stacks
- Additional Combos

Each of these amps have their own sound and dynamic range and when changing their settings make sure that you keep track in the notes section about the changes you are creating.

Pedalboard

Pedalboards are mainly used within the drummer tracks as they will add a different and dynamic range of sound to your drummer tracks. GarageBand offers quite a few pedalboards for use that can be altered anyway you like. Pedalboards go hand-in-hand with amps and while GarageBand offers a variety of Amps, there are only 4 default pedalboards that can be chosen from.

The limited selection might leave you without a sound you like, this means that you can either buy or download a pedalboard from another creator or create your own pedalboard with your own adjustments. There are also pedalboard plug-ins that can be added into the software but those will be covered in the next section.

The four Bass Amps that you can choose from are the following:

- Classic Amp
- Flip Top Amp
- Modern Amp
- Direct Box

When creating your own amp in the amp designer you can just follow the next steps.

- Choose your bass amp out of the list of four above
- Chose a base cabinet from the following available cabinets

  - Modern Cabinet 15"
  - Modern Cabinet 10"
  - Modern Cabinet 6"
  - Classic Cabinet 8x10"
  - Flip Top Cabinet 1x15"
  - Modern 3 Way Cabinet
  - Direct (PowerAmp Out)
  - Direct (PreAmp Out)

PlugIns

Plug-in pedalboards are used to create more diverse sounds when the ones that are used within GarageBand are not sufficient enough for your liking. Most of the plug-ins in GarageBand are created as stompbox pedalboards.

The pedalboard section in GarageBand has the following 3 areas included:

- Pedal Browser
- Pedal Area
- Router

Each of these areas are explained in the first part of Garageband. Along with this GarageBand offers 7 different types of pedals.

- Distortion Pedals
- Pitch Pedals
- Modulation Pedals
- Delay Pedals
- Filter Pedals
- Dynamic Pedals
- Utility Pedals

Alongside the pedals one can also include other plug-ins that allow you different tools and sound changes.

The following are some of the best current plug-ins to include in any project.

- Ambience by Smart Electronix
- MFree FX Bundle
- Synth1 by Diachi Laboratory
- Vinly by iZotope
- Vocal Doubler by iZotope
- TDR Nova
- ValhallaFreqEcho by Valhalla DSP
- Tyrell Nexus 6 by u-he
- Saturation Knob by Soft Tube
- OTT by Xfer Records

These all offer you a dynamic range of changes to any project.

PlugIns Installation Process

This section will outline how to add third-party plug-ins to GarageBand, allowing you to save them and not have to reinstall them with every project. plugins are a dynamic way to add a range of different sounds and effects to any track, whether it be strange sounds or some angelic musical vibes, you will find a plug-in for it.

These will be the steps to follow to install a plug-in to your GarageBand.

- Make sure that your files are Audio Unit formats.
- Download the plugins to your computer, saving them in a file you will remember. Make sure that the download format is MAC AU.
- You can now start the installation process.
- Most plugins will have a plugin installer, if it does not, simply download one that you would like to use.
- Once the plugin has been installed you can move on to the next step.
- Open your GarageBand software, and navigate to the 'Preferences' menu.
- Once the menu opens up, navigate to the Audio/MIDI options and select the menu option 'Enable Audio Units'.
- Now open or load the track that you want to use the plugin with.
- Open the plugin window by clicking on the appropriate icon in the menu bar at the top.
- Find the drop-down menu and search for the plugin name, and select it to start using it on your track and in your project.

Tutorial 3

In the final tutorial we will finish off the project by finalising the last tracks, editing some settings and using some pedalboards and plugins for our track before exporting the file.

Step 1:

I decided that I also want to add some background ambiance to my audiobook so I found some tracks that are free to use on personal projects and downloaded various nature sounds like rain, thunderstorms as well as some fire crackling sounds, and chatter in a tavern.

These will help me set the mood during those stormy sections in the manuscript.

I made sure that the files were all MP3s that I can load into the project as extra tracks. To add them to my project I opened my finder, and the file where I had saved them and simply dragged them across my screen into GarageBand. The software automatically assigned each track their own line and I could drag, cut and paste and adjust the tracks how I see fit.

Step 2:

In this step I followed the editor steps I had in Tutorial 2. I made the fade in and fade out adjustments and then moved on to the smart controls as well as the automation curves. One by one I used the automation curves for volume, panning, reverberation and echoing to create the perfect harmony between all my tracks.

Once this was done I could move on to the next step.

Step 3:

Now I finally moved on to the plugins and pedalboards I used in this project. I only added a few of each since too many might overpower the tracks and end up making the sound seem too muddled to even know what happened.

This tends to happen when the vocals are too soft, or there are too many loud and complex sounds playing at the same time so be careful of anything similar.

Once I had selected the settings for each plugin and pedalboard I could finally move on to the final steps of the project.

Step 4:

I could now select the master track. The master track is the final adjustment phase you go through before exporting your track. It allows you to make any small adjustments to final track volume, tempo, ambiance, and even tone.

Take your time with these settings, save your project constantly and be sure of what you want. Like I suggested earlier, consider duplicating your project and playing around with settings before making your final choice.

Step 5:

Now that you are done and happy with all the choices you have made you can do the last thing. You can either publish your song to Apple Music, SoundCloud, AirDrop, Mail, Export it to a Disk, or even make it available to use with the mobile version of GarageBand, which will be covered in the next chapter.

For this project we're just going to save and export the track to a disk, this will allow you to save it as a track in a folder of your choice. I suggest selecting the high quality default so no information or sound goes missing or gets distorted by the low quality scaling. The export process might take a while.

For the sake of the next chapter, Mobile GarageBand, where I will be covering the mobile version of GarageBand, and how to use it, you can also export it to use it in the mobile version of GarageBand. Simply find the Share menu option and click on 'Project to GarageBand for iOS' and hit enter, you can then save the file wherever you want and just airdrop it to your phone to use in iOS GarageBand. Importing the track to your mobile version of GarageBand will be covered in the next chapter.

MOBILE GARAGEBAND

Mobile GarageBand is absolutely fantastic and super easy to use. The user interface is slightly different, as most mobile apps are different from their desktop or laptop counterparts but it is ultimately easier to use. In this chapter I will be discussing and working through the software with you by using the track we created in the previous three chapters.

What makes the mobile version of GarageBand even easier is the fact that you can plug your guitar and/or keyboard directly into your phone when recording the tracks. It also allows you to edit and change tracks on the go. I absolutely adore the way the mobile version is created.

I will cover the basic layout and design as well as discuss how to use the software to its fullest potential.

Basic Interface

All Apple devices automatically come with GarageBand preinstalled, but let's say you, like me, value space above anything else you might have deleted the app since you don't use the mobile version often. I generally only ever use the mobile version when I'm traveling and working on a project. Generally I will download the software onto my mobile device and from there move the tracks I need from my Macbook or Mac via AirDrop to my phone. This allows me to save some space on my phone, even though I have to go through the hassle of re-downloading the app again.

The easiest way to do this is simply to go into the App Store, hit the search bar and type in GarageBand. While there are multiple different products with similar names, make sure that when you select the app it is the correct GarageBand as downloading the wrong software might cause you to lose your work, or destroy your track and stop you from loading it back into the desktop version of GarageBand later when you want to edit or adjust settings.

The download might take a few minutes, depending on your network as well as the internet speed. The iOS GarageBand app is rather large, taking up about 1.6 gigs worth of space. Before downloading the app I suggest taking the time and making sure that you have time and space to download the software. The image below will show you what the app looks like in the App Store.

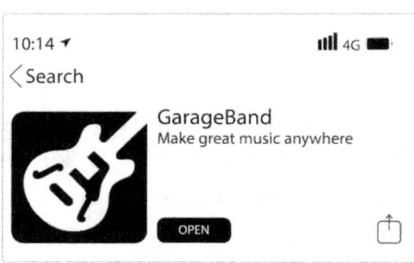

Now that you have downloaded the app you can finally open it. The first time you open up the app you will get various prompts asking for access to your microphone as well as all the new updates that prompt if you want to allow the app to track your data. You can select the settings as you see fit and how you choose to use these settings.

Once this selection has been made you will be prompted surrounding all the changes and extra data that you might need to download later. You can choose to ignore this but like I mentioned with desktop GarageBand, you will have a limited library and you might end up downloading the extras anyway. The choice will stay yours though.

Once you have gone through all the prompts the first screen will pop-uo. Since the mobile version does not have a mouse, you will be doing a lot of sliding and tapping with the software.

As with desktop GarageBand you will have to select what type of track you want to create. There are a lot more options in the mobile version of GarageBand which is why I like it so much.

Instead of selecting between an audio, software instrument or drummer track you can now select between 11 different options.

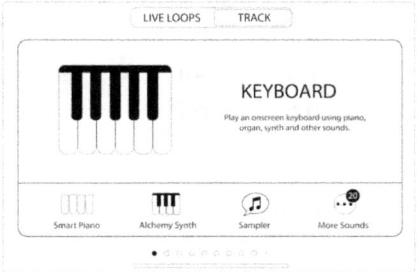

GarageBand will automatically open up to the keyboard screen. As you can see there are scrolling dots at the bottom, indicating the scrolling selection that can be made.

The following is a list of all the options that you can select from when selecting a track type.

- Keyboard
- Drums
- Amp
- Audio Recorder
- Strings
- Bass
- Guitar
- World
- Drummer
- External
- Sound Library

Each of these selections have a different track and track settings, allowing you to create an absolute unique sound no matter what you are looking to create.

In the following section I will cover each of these menu options and how you can use them inside your project.

Keyboard

The Keyboard track lets you select between three different track types as well as offers you a fourth option to scroll through the library to find a sound you would like to test or use within your project. The Keyboard track selection creates a digital keyboard that you can play on your phone while recording it, using the software.

Smart piano lets you select a piano that will keep tune and pitch while you play, correcting any mistakes as you go on, also having predetermined settings that will make the usage of this selection the easiest.

Alchemy Synth creates a track that sounds more electric and sounds more like an electric keyboard that should go into a pop or dance song than an orchestral feel. You can however use it as you see fit.

Sampler let's you sample a track or at least a small selection of a track, even one that might be your favorite song right now. This is an amazing tool if you wanna use another song as inspiration for your own. Remember that recreating a sound exactly is a copyright infringement and it is best to create unique sounds that are inspired by favorites instead of copying and stealing.

The final menu option that you can select would be the More Sounds menu item. This allows you to open up the library of keyboard sounds and scroll through them. This is especially handy if you aren't proficient in keyboard or piano playing and would like to add something like an organ or synth player into the mix.

Drums

This menu option might seem confusing as there is also a Drummer track selection that can be made via the track selection process. This selection let's you select acoustic or electronic drums of your own, making the entire process far more creative and authentic than using a preformulated and recorded track or loop.

The first selection is Smart Drums, this selection has pre-set settings that will allow you to simply hit the record button and tap the cymbals, percussion sections and the hi-hat or treble bass. This makes the recording process far more fun as it allows you to play around with the sound while not having to worry about the settings.

Secondly you have the Acoustic Drum selection that can be made. This creates an acoustic drum set that you can play in the same way as the other drum set, letting you focus on the sound while the software does the heavy lifting for you.

The third selection, called Beat Sequencer, creates a large square with smaller squares inside, each square linked to a beat sound that can be recreated simply by tapping on the corresponding square. These are great for pop-songs and EDM music, but you can also use them within the normal projects if you just want to create a recurring sound throughout the track. They are fantastic for creating your own loops.

The final selection you can make of course is the More Sounds one, leading you to the library to create easier access to more sounds that cannot be found within the three previous selections.

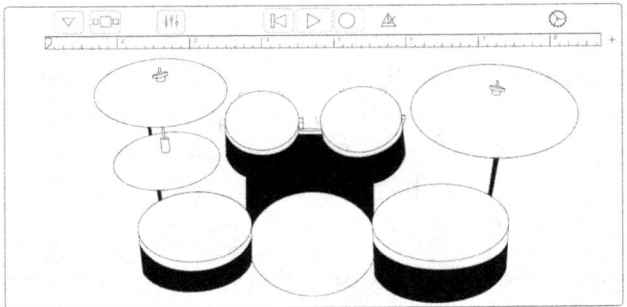

The above image shows what the app looks like once you have selected the drums or even the drummer selection. Here you can tap on any of the objects and it will reproduce the sound for you.

Amp

The Amp selection takes over the work of the Amp Designer in desktop GarageBand, letting you select the type of amp you would like to use in your project. Again there are three different selection types and a fourth that will lead you through the library if you cannot find what you are looking for.

The first selection would be a Clean amp. This selection adds a clean amp sound to your project, letting you adjust the settings however you want in the track screen and creating clear cut sounds throughout the project.

The second amp that you can select is the Distorted one. This amp does exactly what the name implies. It creates distortion within the track, whether you are using an acoustic or bass guitar.

The third selection method is the Bass amp. This amp is perfect when recording a bass guitar sound, making it easier to create those deep bass sounds that any project needs.

Audio Recorder

The Audio Recorder selection will be the one you select when you want to record some vocals or audio that would not be a typical instrument, but you can opt to record an instrument here.

The first selection would be Audio, letting you record any voice prompts or sounds, while the second, which is Instrument, allows you to plug your guitar directly into your phone and play it, letting the software do the recording work for you. Make sure that when you do this that you use the correct wiring and auxiliary cables as too much power and amperium might damage your phone.

Like with the previous menus you can also opt to go through the library to find something a little more to your liking.

Strings

The next track selection that you can make is the strings collection. While most people think of guitars when thinking of stringed instruments, you can also play and record instruments like harps or violins. This menu option is mainly created so you can play and record orchestral or solo string instruments. The three options here are a little more complex, but nothing you can not play around and learn with a few minutes of playing around.

The first selection will be Smart Strings, this allows you to focus on playing your instrument as perfectly as possible and letting the software tune and tone your playing to match the project you are working on.

The second selection is Notes, this will show you the notes while you are recording, allowing you to make a note selection once the string instrument has loaded, letting you better create a sound that is linked to a certain note.

The third selection you can make, besides the library selection, is the Scales selection. Like with the Notes selection you can select the scales you want to record in, making the project easier to bring together and flow.

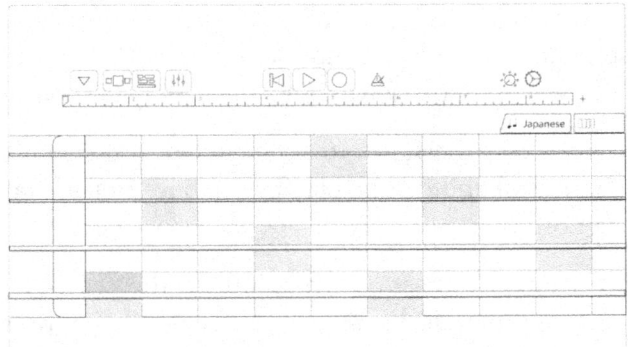

The above image pictures the string instrument as well as the Genre menu option where you get to select the genre, currently set to 'Japanese' and the icon next to it where you have note selection.

Bass

Next up we have the Bass track option. Of course this relates to Bass guitars and like the other track selections there are three selections excluding the library search functionality. These options are very similar to the string instrument menus and do similar things.

The first menu is Smart Bass, letting you focus on the playing part while the software does all the challenging things like correcting tone and small hiccups.

The second menu Notes, again letting you select the type of note as well as the genre, allowing you to get as specific as you need to.

The third menu Scales, lets you select the scales you want to record within, discarding everything else.

Guitar

This track selection is the easiest to explain. Most tracks have guitars in them, even if it doesn't always sound that way. Like the Bass and String tracks the Guitar track has three basic track types excluding the option to search through the library for what you have been looking for.

Smart Guitar, like the rest of the other Smart instrument selections, helps you focus on playing the track you want to record while doing all the editing and adjusting within the software. You can also adjust the track afterwards if you don't completely enjoy the sound.

Notes lets you select the notes to be included and excluded in the project as well as aligning your playing to those notes, making sure you are never off-key.

Scales allows you to adjust and manipulate the scales, while also making sure that the project is recording as smoothly as possible.

World

Of all the different types of tracks that the iOS version of Garage-Band has to offer I think this might be my favorite track type. The World track type allows you to play different instruments from around the world, without having to learn how to play them.

The first selection is called Pipa, which includes stringed instruments that helps you play the instrument without having to worry about the settings and setup.

The second track selection is called Erhu, letting you play instruments similar to the Erhu sound.

The library of course offers far more instruments to choose from, and will allow you to broaden your horizons when it comes to different types of instruments.

Drummer

The drummer track selection is similar to the drummer tracks that can be found in the MacOS GarageBand version. These tracks are pre-recorded drummer tracks that can be manipulated and styled the way you want them.

There are two different track selections excluding the library drummer selection.

The first selection you can make will be Acoustic, letting you choose from a grouping of different acoustic drummer sounds.

The second selection is Electronic, letting you choose between a variety of electronic drum recording and sounds.

External

This track selection allows you to use plug-in tracks or to plug your instrument directly into your phone. Doing this allows you to play your track and record it on your phone. You do need to make sure that you have all the correct audio equipment for this as well as the correct audio jacks to connect your phone to your guitar, drums or keyboard.

Remember that GarageBand uses Audio Units so it would be a wise decision to invest in an Audio Unit Extension for whatever instrument you use or play.

If you do not play an instrument and would like to record a friend or colleague make sure that they have the correct cables and plugins for this to work. For guitars and keyboards I would suggest getting a MIDI connector that relays the information to the Audio Unit Extension.

Sound Library

The final menu option that you can select is the Sound Library one. This is basically the entire library where you can pick any track you would like to use. Here you will select from a wide variety of instruments, songs and various loops to use in your project.

Simply select the sound library options and a large library section will open where you can choose from a variety of different instrument packs as well as professional artist remix sessions and loops.

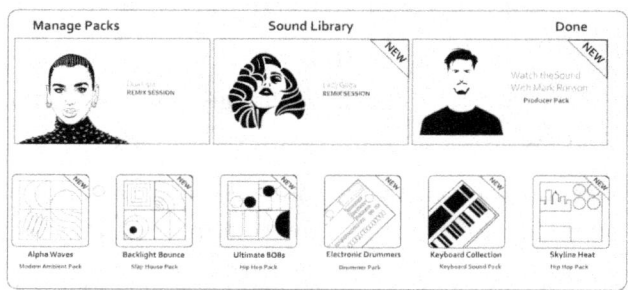

These packs will include different sounds and settings and there are hundreds and thousands of different sounds and packs to choose from. Whether you are looking for some basic keyboard sounds to 70's music that was played once on that album that no one really remembers you will find the sound you want for your project. Remember that these will have to be downloaded so make sure you have good network connection and space to do so.

Basic Functionality

Now that we have gone over the basic starting menu for the iOS version of GarageBand we can go over the basic functionality and how the iOS version differs from the Mac version. As stated before the interface is somewhat different and more compact so I will do my best to add as much detail as possible to make it easy to understand and consume.

At the end of this chapter I will include a standalone tutorial for how to import the file we worked on earlier in the volume to your iOS GarageBand and how to use it for your projects, so keep an eye out for that section.

Once you have selected your track, no matter what it will be you will be greeted with the corresponding screen. For user purposes I will select an audio track so explain the basic interface. I will start with the header bar.

Starting from the left hand side I have used colored blocks to indicate the different buttons that you will find. Later in the section I will show you the header bar for when you open the tracks section as it looks somewhat different.

- **Red**: The upside down arrowhead icon indicates the track information that is being hidden. It will show you the library option of selection tracks from your library or whichever genre or instrument you have selected.
- **Yellow**: The three screens icon is similar to a back button, taking you back to the track selection page where you can choose a new track.
- **Orange**: The different lines icon indicates the track section, when you press this icon it will open the track view, allowing you to see your tracks the way you would see it in MacOS GarageBand.
- **Green**: The adjuster icons indicate the changes you can make to your tracks. This will show the small settings that you found on the left hand of the MacOS GarageBand where track volume, compression and Master Effects are indicated.
- **Turquoise**: The turquoise section is the recording LCD section. The first icon is a rewind button, the second a play

and pause button and the final red dot is the recording button.
- **Blue**: The Triangle icon is also found within MacOS GarageBand as it is the metronome that keeps track of you while you play and record.
- **Pink**: The backwards pointing arrow icon is the Undo button, this allows you to undo track creation or setting changes.
- **Purple**: The purple icon is the settings for audio recording. Tapping this icon will open the audio recording settings that change information on the channeling for the track, the tone and squeeze as well as the Monitoring. This is important to note as it will change to the Loops icon, as seen in the next header, that hosts the loops.
- **White**: The final icon is, of course, the settings icon. This opens the settings menu with regards to things such as tempo, key signatures, time rulers, fading in and out, the note pad section and any advanced settings you might need in the future. You can also switch on the count-in that is found in MacOS GarageBand, letting the software count in for you so you can focus on playing the track you want to record.

Now these menu options change when you start recording sound, most of these menu items stay the same but there are some changes.

In the header bar as pictures above there are three major changes to the icons and functionality.

The Track selection icon changes (yellow) to a singular track icon. This purely means that you can now adjust that track and recorded section as you like, dragging it to make it longer or moving it around inside the track.

The second change that we come across in the menu is the new FX icon (red). Pressing this button will open the effects section at the bottom of the screen, allowing you to add various effects to your track.

The third change is the loops library (orange). This will open the loops library for all pre-recorded loops as well as any extras you may download from the sound library or may pre-record on MacOS or iOS GarageBand.

It is important to note that all of the buttons in the iOS GarageBand can be tapped a second time to close any menu or screen section that opened up when you pressed them the first time.

Also be aware that depending on which track selection you make that the Yellow icon in the second header will change to the corresponding track. For example, that singular track icon goes with drummer tracks, whereas audio tracks will have a small microphone icon that will take you to the audio recording screen to adapt your settings.

Tutorial

In this tutorial I will be guiding you step-by-step through the process from beginning to end on creating a track from scratch in the iOS GarageBand mobile app.

I will be as thorough as possible and include as much information as possible. I will also be adding some hints and tricks for when it comes to editing, track creation and the best way to use the software for your project. Some steps might seem simple while others might seem more complex. I will be breaking up the complex parts into multiple sections.

Step 1:

The easiest way to start is to sit down and make notes of what you want to create. Now that I have finished the first chapter for my audiobook project I know that I have about 12 chapters left. My idea is to record each chapter separately, then slowly add in all the background sounds and soundtracks before adding the 13 tracks into one file and creating one long audio file.

Creating one long audio file will make it far easier to listen to. I considered keeping the chapters separate as some people may enjoy finishing one chapter at a time and for that reason I will be keeping duplicates of each chapter. As I mentioned before, always keep duplicates of your projects, and be sure to save them to your cloud drive. This is quick and easy to do and keeps you from losing valuable information or the entire track.

For my second chapter, I reviewed the entire chapter and I know that there are no singing sections in this specific chapter, but there are scenes where the characters are sitting by a fire in the woods. So instead of making notes separately in a notebook I am going to use the note pad functionality in the iOS GarageBand version. To do this, simply open your GarageBand on your phone, use the top left-hand upside down corner icon and select 'My Songs', click the plus icon in the top right corner and create a new track.

This will take you to the main screen where you can now select a new track type. Since I will be mainly recording audio, I will opt for the Audio Recorder selection, with an audio track type. This will open the following screen.

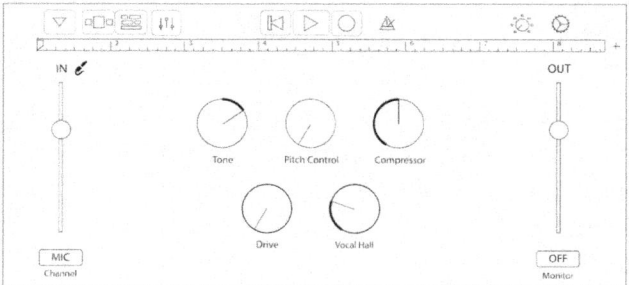

Once you have opened this screen you can simply select the settings wheel in the far right hand corner. This will open the entire track's settings menu. Scroll down to the note pad section and simply press on the little arrow to open the note pad section.

The note pad section will look exactly the way it looks in the note pad app on your iPhone. Simply tap the blank canvas and make the notes you need to make, and start typing. Once you are done you can simply hit the back button and you will be able to go back to the track area where you can start recording.

Step 2:

Now that I have notes on what I want my project to be and I know how I want to record the track I can move on to the recording process. This may take a while to get used to, and I urge you to take your time. Make sure you have all the correct connections from your mic to your phone, whether you are using your headset or a professional microphone setup, be sure to have the correct things plugged in.

Generally I will record a small section of me talking, and just testing the mic. This allows me to gauge how loud or soft I need to talk into my mic. When using a professional mix setup this is easier to control as most microphones have a volume adjuster on the base. When you are using your headset from your phone you will have to either move your mic away from your mouth to keep it from creating a muffled sound or move it closer to keep the sound from being too soft. I generally use the following samples to test my microphone sound. While there are a few, I have experienced that using all of them allows me to accurately gauge if my microphone is working or not.

I found these sentences that are also used by the military are the best way to test my microphone in every capability. (Reed, G. n.d)

- The beauty of the view stunned the young boy.
- It snowed, rained, and hailed the same morning.
- The small pup gnawed a hole in the sock.
- Two blue fish swam in the tank.
- Read verse out loud for pleasure.

Simply repeating these lines into my microphone helps me test every type of sound.

Step 3:

Once I have recorded myself repeating these lines a few times, in a random order, I will listen to the playback, both with my headset and without my headset. This really gives me the capability to tell if certain sounds will get lost if I add other tracks or sounds later on in the project.

Once I am happy with the playback on the normal recording I will go into the settings of the screen that I showed in step one. I will adjust one knob at a time, using the note pad section to keep track of all the changes that I am making as I go along. With each adjustment I will record myself saying one or two of the lines mentioned above, or even just rambling about my favorite topic. Sometimes it is easier to tell if you are rushing through words when you are talking about something you enjoy discussing with someone.

I will listen to the playback and once I am 100 percent happy with how the playback sounds I will start the recording process. I will delete all the test tracks that have been recorded and then start.

Step 4:

When you record, it is important to let there be a moment of silence. This allows you to sample from the two or three seconds if there happens to be some background noise later in the track.

Do not be afraid to re-record sections. This happens to all of us. I would suggest taking a few seconds and just starting the paragraph of chorus from the top. Simply let a few seconds of silence fill the audio track and then start again. This will help you, especially since you won't have to fumble with pressing the record button twice. If, however, you do want to use a new track and start off where you fumbled you are more than welcome to do so.

Simply tap and hold the audio track and when the menu pops up, select 'duplicate track'. This will duplicate the settings and the current audio of the track, to use the second track, simply delete the recorded section and once you are ready just press the record button, wait a few seconds and start recording.

Step 5:

Once you have recorded all you have recorded it is time to listen to all the tracks that you have recorded. This might be a jamble as all of them might start at the same time if you opted to use multiple track recording, if like me, you chose to record one long section your step five will be somewhat different. I will cover both options.

Option A: if you use multiple tracks, I would suggest muting all the tracks except the first. In the iOS GarageBand version you can mute tracks simply by hitting the adjuster icon at the top and muting the track once the track settings show up on the right hand side. This will allow you to listen only to the tracks that have not been muted. This will make editing far easier as you will only be working with one track at a time.

Option B: if you, like me, used one long track, you will have to be far more strategic than using the first option. When going through the playback you will have to select the sections where there was faltering and problems and cut that section from the track and then shift the two track sections back together. This takes far more skill than one thinks so if this is your first time using iOS GarageBand I would suggest opting for option A.

Both of these options will take you a while to complete. Take your time to go through each and every track, making sure to cut out any unnecessary sections and adding all the effects, as well as fade in and fade out. This is important as it will allow your track to flow together far smoother than it originally did.

Step 6:

Now that you have done the basic editing for your track you can finally start moving on to adding the other sections of your track. If you are adding any of the track selections that you can find from within the 11 selections I will be covering how to create those tracks in this section. As I already covered the audio tracks, and the sound library I will only cover the other 9 track selections. As with the audio tracks I suggest playing around with all the tracks and their settings to create the sound you want to add to your project. There are no limitations and you can add as many tracks as you want, even if it means having 9 different drum tracks. This is your project and you can do whatever you want.

Keyboard: Simply add the selected track selection, whether it is a Smart Piano track or an Alchemy synth, click on the appropriate track and add it into the track selection. The software will automatically take you to the controls section of this instrument. Use the LCD above to start recording. Before starting to record, get used to the user interface and the way the sound works before you start recording. I find it easier to create a sound once I have created a few derpy sounds that have not been recorded. Once you are comfortable with the controls, record and edit to your heart's content.

Drums: Add the desired track selection and play around with the controls, make sure you double check the menu options to be able to navigate the software a little easier but again, play around with the controls and settings and then dip your toes into the wonderful world of drum tracks.

Amp: Ampos are used to create deeper and broader sounds, get used to the settings and controls and adjust them as you need. Record your sound, adjust after playback, make your notes and then move on to the next track selection.

Strings: String instruments look a little more different, creating a 3D image of the string instrument to allow you easier access to play the track. Make sure you get used to the controls, play around with the note and scale settings and then rock out with your favorite sounds.

Guitar: Guitar tracks are similar to string tracks as they also create a 3D view of the instrument, letting you play the strings as you would a normal guitar, I know what's helped me is holding my phone like I would a guitar for just a second to let me get used to how the chords are structured and how they play.

World: The world track is a little different than the normal tracks as each instrument would be set up differently. Take your time to get comfortable with the controls and then dive in head first.

Drummer: The drummer track, like the drummer track in MacOS allows you the same 3D view of the drums, only from the top, and instead of clicking to create the sounds you will simply be tapping on the different parts of the drum set to create these sounds. If you opt for the Electronic track you simply will be tapping a small square that will light up and create a sound for you, making it even easier to use.

External: External tracks are for tracks that need to be added into iOS GarageBand instead of using tracks from the library or your music app. In the next step I will explain how to import different files into iOS GarageBand.

Now that you have added all your tracks you can go forth and create wonderful melodies and harmonies and distortions.

Step 7:

In this step I will cover how to import tracks from elsewhere into iOS GarageBand. As mentioned before we will be using the track that we created in the earlier tutorials for this purpose.

When you export your track using the 'Project to GarageBand for iOS' selection the file was saved as a 'mobile.band' file type. This is important as Mac GarageBand changes the file type automatically for you to easily import into your mobile version of GarageBand.

So you have the file on your computer, and you have to move it over. There are multiple ways to do this. One way is to connect your iPhone to your Macbook or Mac desktop with the lightning cable and then move the file over. While this is easy I generally opt for the second method. The second method is using AirDrop.

To AirDrop a file simply switch on the WiFi and Bluetooth on both your Mac and iPhone. Make sure you select the track in your finder and right-click. The menu option to to share will pop-up, click on it and then select AirDrop. Click it. Your Mac might take a while to find your phone if you have never done this, however if you have done this your computer will quickly find your phone and you can just click on your picture and your phone will beep letting you know that a file transfer is happening.

Your iPhone will prompt you, asking where to open the file. Garage-Band will be the first bolded option, simply click on GarageBand and it will open and load the track into GarageBand for you.

Now comes the second part of this section. Now that you have imported your song into GarageBand you can simply click on the file and delve deep into the editing process. Once you are done editing the track go back to the 'My Songs' section. Here you can share the file to become a ringtone, a song or a project. If you want tore-use the track back in Mac GarageBand you can simply select Project and follow the prompts to make the file available for your desktop editing later.

Step 8:

Now we can finally move on to the editing of the tracks using the effects section of mobile GarageBand.

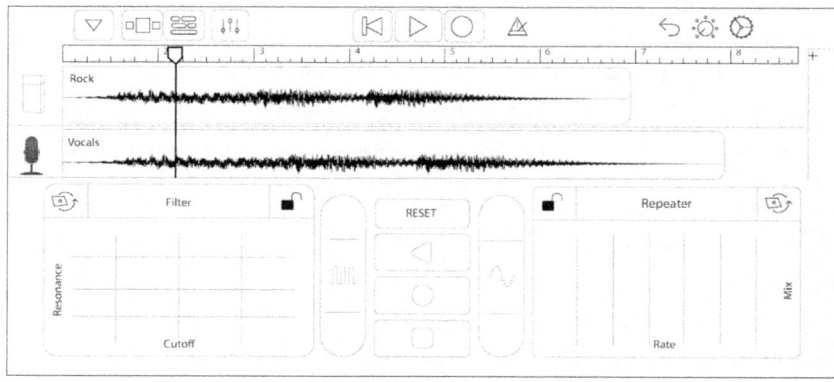

The effects panel in GarageBand looks like this, and as you can notice in the top LCD screen the FX is a pink color, indicating that the FXC screen is in fact open and can be used for any of the tracks.

As you can see GarageBand automatically turns the selected track blue, which is vastly different from Mac GarageBand since it uses color coding to keep tracks different.

Now that you have finished most of the work you can start editing the tracks. I can give you a bunch of information when it comes to editing and what some people do but in the end it is entirely up to you so the best way to get through this phase is to play around with the editing buttons and to see what it does. This will help you better understand how the effects work and what their impact is on each track.

Step 9:

Now that you have finished all your editing you can simply export your track to the song selection. This will allow you to save it in your Apple Music library and allow you to play it for everyone. Creating a ringtone with your project will allow you to use the track as a ringtone later on.

Selecting the Project option will export the file so you can use it as a file in Garageband, this is super useful as it will allow you to work on tracks when you are back at your computer and it needs the final fine tuning.

You do not need to use Mac GarageBand to finalise your projects. I always opt to fine tune and double check my files before exporting them to the appropriate format as I might miss some minor details some days.

Tips and Tricks

In this section I am going to give some handy tips and tricks for iOS Garageband and how to get the most out of your usage with the mobile software.

Custom Chords

By selecting the Guitar and the Smart Guitar selection you can add your own custom chords to any track. Once the main screen has opened after you made your selection, simply tap on the settings icon in the top right corner and then select the 'Edit Chords' menu item. This will allow you to change the chord strip that is automatically shown at the top of each chord and filter through the chords given to make your change. You can also add an alternate bass note by swiping up and down on bass selection. Then simply hit the 'Done' button in the right hand corner to finalise the change.

Keyboard Note Labels

iOS GarageBand allows you to add note labels to the keyboard and piano keys in the app. Unfortunately it is not just a software issue but you will need to change some settings on your device as well. To do this you will have to open your device settings, and navigate to the GarageBand app list at the bottom and open it.

Simply scroll down in the menu until you find the 'Keyboard Note Labels' and slide the toggle on. Once this has been done, restart your app and you will now be able to see the note labels on the keyboard and piano tracks in GarageBand.

Acoustic Drum Patterns

You can have any of the acoustic drum sets play a repeating sound over and over simply by pressing and holding down two fingers over that selection, whether it is the base or the cymbals or even other selections of the percussion and kick-drums.

Adjusting the space between your fingers on the selection will either slow the repetition down or fasten it, and moving your fingers up or down will either decrease or increase the volume of the playbacks respectively.

Record Into Live Loop Cells

GarageBand allows you to create live loops on the go. What makes this amazing is the fact that you can record into any of the cells within the Live Loops grid.

Simply tap the cell where you want to record your live loop inside of and hold down until the menu to 'Record into Cell' pops up or simply just double-tap the selected cell and GarageBand will know that you want to record into that cell.

If a touch instrument has been linked to that cell, double tapping the cell will automatically open the instrument and allow you to play and record it. However if a real instrument was assigned to the cell the audio recorder selection will be opened and you will be able to record and replay the selection on your real instrument.

Once you hit the back key the cell will be saved and the entire project will be saved as a new project that can be viewed and edited later on.

Custom Drum Effects

GarageBand allows you to create your own electronic drum kits. You can edit a basic electronic drum kit by changing the drive, crush, low cut and high cut controls at the top of the screen. Once you have finalised your edits, grace the kit with a new name simply by tapping the name in the middle and renaming it to whatever you would like. To find and use your custom electronic drum kit again search in the custom drum category and you'll be able to use it for any project.

Plugins

Plugins allow you a wide variety of changes and edits and will create sounds and effects that you didn't even know you had. New plugins are being released daily and keeping up with all the different types and what they can do for you and your project is exhausting. While there are hundreds of different plugins that work wonderfully, I would suggest the following plugins.

- Flux Mini
- Wider
- LCR5
- Rough Rider 3
- DLYM

These are some fantastic plugins and will allow you to create almost any type of sound you are looking for. Feel free to search up various other plugins and use those to your heart's content. I enjoy using these in my projects.

Be sure that when you are looking at plugins that they are iOS and Mac compatible. Always double check that the download is from a secure website and make sure to run antivirus when you are downloading from the internet continuously.

USB Audio Interface

There are various Audio Interfaces that can be used along with your iPhone and iPad, but the model and the capabilities all depend on the type of project you are planning on working on as well as your skill level. I think the best way to gauge how important or needed this is, is to assess your skill level, the things you need and then go from there.

I also suggest doing as much research as possible, making sure that when you choose an audio interface device that it suits your needs and skill level. As much as I want to say that you can simply plug-and-play, it seems to never be that simple no matter how hard you try. In this tips and tricks section I will be covering how to use your audio interface as well as any minor troubleshooting you might come across when your interface is not working.

To start off, audio interfaces can be plugged into your iPhone or iPad using different cables. You get lighting cables, which are the common terms used to talk about Apple products, these allow the information to flow through the cable at a faster pace than just perhaps a normal cable. While you can use lightning cables you can also opt to use USB-C cables for this, but no matter which cable selection you go with, the process will stay the same no matter what. It is important to note that FireWire interface cannot be attached to iPhones or iPads.

While you can use older models, I will be covering the connection process for an iPhone 5 and up as well as an iPad 4th generation and newer as these use the lightning jacks instead of the older charging ports. You will first have to buy yourself a lighting to USB-adapter, this will allow you to plug the USB-C cable from the interface into the adapter and the adapter into your phone, using the opposite end that is the lightning port. The adapter is also known as the 'Camera Connection Kit'.

While it is best to use official Apple products, you are more than welcome to use cheaper alternatives from Amazon or Ebay. The problem with not buying an official apple product is that they are constantly sending out software updates that might make your non-apple hardware stop working, which means you can no longer use the item. While Apple products are a little more expensive, I would highly advise on getting the Apple official product, for your safety and the ease of use later on.

Make sure that all the lightning ports are cleaned and dust free and then plug the port into your device. Do not connect the adapter to the audio interface first. Plug in the audio interface into the adapter, make sure that you are using the correct cable and that you are plugging in the USB-C into the correct port, in the correct manner. The first problem you will encounter will be a pop-up that states that the audio interface needs too much power, and that is because neither an iPhone or an iPad create enough power to power your audio interface.

This is why the official Apple adapter comes in handy. The adapter also has an extra slot, next to the USB-C slot for another charging port. This means that you can plug your charger into the port, giving the audio interface and the iPad or iPhone enough power to be able to run the interface. This means that you will now be able to record your audio on GarageBand.

The iPad Pro is the most user friendly in this section as it outputs enough power to let your audio interface run without having to plug in anything else. It is also outfitted with a USB-C port instead of a normal lighting jack like the iPhones and older model iPads.

Guitar Recording in iOS GarageBand

While recording with your built-in microphone is the easiest and quickest way to get some instrument sound recorded it is something I highly suggest not doing. Background noise, natural ambience and even the slightest movement will create distortion on your track that you do not want. There are various other ways to work around this. I would suggest using this method when working on demos or you want to create a low frequency sound.

One of the older methods was to use the iRig audio interface and while this worked wonderfully for a few years, this method left the building once iPhone passed the normal earphone jack and went to lightning jacks instead. The new updated first generation HD iRig can now be connected via the lighting jack on an iPhone, allowing you to record your guitar tracks.

Audio Interfaces are also an amazing way to do this. As this was covered in the previous section I will not be going over that again. Feel free to play around with various interfaces and tools to create the sound you would like to create.

Compression for Beginners

This is one of the most useful effects that any single person can use, no matter what your skill level and track type, using Compressor will create a fantastic balance for all the tracks in your project. This is why I am covering this specific effect in detail for you. I use Compression all the time, allowing me to tone down the loud explosive sounds that oftentimes sneak into my projects.

Compression is a fantastic tool to use on all your individual tracks as well as your master track in the end. Compression lowers volume on harsher, louder sounds to the same level as the quiet sounds that can be found within your track. Once this has been done it is always important to alter and adjust your Gain as Compression might lower peaks and troughs in your track.

It is important to remember that the amount of compression used on any specific track is heavily dependent on what type of instrument you are playing as well as what the music genre is of your track. Generally the rule of thumb would be to use less compression on instruments like Violins and orchestral sounds and more compression on Hard Rock instruments and tracks.

Overcompression is something that a lot of beginners struggle with, especially in the mobile version of GarageBand. Over compressing a track will leave the track sounding dull and lifeless.

The compression settings can be found in the general settings area. To toggle this, simply drag the track icons to the right of your screen and tap the adjuster icon in the LCD. This will open all the basic effects that can be used. The best way to use the compressor effect is not by using the single slider, but instead using the entire interface to do your compression, this will give you better tools and adjustment to keep you from over compressing.

The compressor setting has 5 different controls.

- **Threshold**: This setting controls the point when the compression starts and lowers the volume of loud intrusive sounds. This slider works in decibel levels and using it will soften any sound with a higher decibel level will be toned down.
- **Ratio**: The reaction will be the determinant factor on how much of the track volume you want to reduce. A higher ratio will end up reducing more sound and a lower ratio will reduce less volume. For example, a 2:1 ratio will be read in the following way; for every single decibel that goes over the threshold the volume will be turned down by two decibels.
- **Attack**: This setting will be in charge of the speed rate at which the compression will start in the project. The attack

speed will vary depending on the sound that you are compressing. For example, instruments like Hip-Hop, snare drums and kick drums will traditionally be placed on a fast attack setting whilst guitars and pianos, or instruments that need to sound more natural, will have a slower attack setting.
- **Gain**: Increasing the gain will push the output level higher, this is important as your track might sound lifeless now that all the tracks are playing at the same note volume. To do so, you will need to slide your gain control higher than before.
- **Mix**: This setting allows you to adjust the amount of compression that you will actually hear in the track. For example, the most common use is it set Mix to a 100 percent, this will only output ot play the audio where the compression has been added, so using the mix slider on the 50 percent marker will allow you to hear the entirety of your track without losing quality or sound, as it will mix the compression track and the normal track together in perfect harmony.

What makes the compression in iOS GarageBand so amazing is that the software automatically adds a compression preset to your tracks. While this will be fine if you are just starting out, I suggest going into the settings and tweaking them a little to find what works perfectly to you. Each track and creator will want a different sound so the settings are all ultimately up to you and if you wish to not use compression then that is perfectly fine as well.

Autotune

While recording yourself feels great, it isn't always the best feeling when you have to listen to the playback. There is absolutely no shame in using autotune and it has become a staple in almost all music that can be found nowadays. In this tip and trick section I will cover the basics of using autotune in the iOS GarageBand app.

While autotune can be used to completely alter someone's voice, you can also nudge someone's amazing performance to be a little tighter than it was, making sure that the track comes out amazingly. While you can crank the settings up to create a robotic sound, I suggest using a light touch. Like some of the effects and settings found in the mobile version of GarageBand, getting used to these settings and what they can do, will and can take a while, so take your time to play around and see how the changes you make affect the track you recorded. This will be the easiest way to learn exactly what these controls do and how they impact your tracks.

While pitch correction is a fairly easy and straightforward process, I do suggest taking the time to really listen and engage with your track, making sure that you are not too heavy handed about the effect.

The first and most important step for using pitch correction is to make sure you know what the key is that you are working in. Selecting the wrong key will autotune the sound to that key and you may end up with some horror vocals instead of that angelic ballad you were hoping for. One of the ways to identify the key is by using the first note in the scales and making that your starting point. There are however different ways to find your project key.

I suggest using a website called Audiokeychain. Here you upload your track, after exporting it of course, and then let the online application do its job. The online app will estimate the tempo and the key of your project. While this method is not foolproof I have not yet found a reason not to use it. If you are more musically inclined than I am, you can simply go off by ear, simply by opening the grand piano track, and pressing the keys until you find something that sounds similar or identical enough to your track.

Make sure that you know if the key is in major or minor, as this will have a great impact on your track and pitch tuning later. To identify if the key is in major or minor, simply play the note a major third up from the note you have found to be your tonal note. If it fits with the song as well you can assume that the project key is C Major.

You will need to change the key of your track to the correct key before starting the auto tune correction. To do so, use the settings icon in the right hand corner of the LCD to open the settings menu. Tap on Key Signature and adjust the key signature to the key you have found your project to be on, tap the save button and your key has been set up. GarageBand always uses the default that is C Major.

GarageBand also has a toggle in the key menu called 'Follow Song Key'. This will change the key for each track in your project and adjust the sound of those tracks. This will not affect the audio recordings that have been done and included in the project. I would suggest finding the key of your project before recording any audio tracks, just to make the process a little easier for yourself later on.

Now that that has been set you can move on to finally starting the pitch correction process. If these settings have been done you can either apply them to your tracks now or re-record the audio tracks and have the pitch correction automatically added to your track as you record. Let's assume that, like me, the idea of re-recording various audio tracks is a lot of extra work, so I am going to quickly guide you through the process of adding tuning to your already recorded tracks.

Select the first track you want to work on and tap the microphone icon in the top left LCD screen. Most of the audio tracks have pitch correction built into it, so it's up to you to activate and adjust these settings to make the audio sound fantastic. Slowly turn the knob, letting pitch control do its thing. The best way to do this is to turn off all the settings except the pitch control and then adjust to your liking. Continue playing around with all the settings for each track until you are happy and add in the effects you would like to use.

This will allow your track to sound far more professional than just hitting a few buttons and claiming you are now a music genius.

USB Microphone

While recording with your basic built-in mic is absolutely fine and you can get away with it to some extent, I do suggest always trying to record your tracks and audio with a professional mic. As with the

Audio Interface, you will be needing an adapter to enable you to plug your mic into your phone and allow you to record.

The easiest microphone set up is to opt for a USB microphone. There are quite a variety of microphones on the market right now and while I will always suggest quality above all else, I also suggest shopping around. If you are only going to record a few tracks every few months or so, investing in a mic that is $400 and upwards seems a bit excessive, so find something that fits both your pocket and future goals. If, however, you will start recording endlessly and constantly I would suggest saving up for a great quality mic. It is fine that you opt for a cheaper one in the beginning but if your main goal is to start recording vocals then having a more expensive mic will help you out in the long run, both in terms of the quality of sound you are producing and the length of time your mic will last before having to upgrade or replace it.

What makes the decision on what mic to get even more difficult is the fact that you get different mics for different types of recording. There are designated mics for ASMR, and streaming or video creation on YouTube, as well as more basic mics that can be attached to your clothing for Blogging. It all just depends on the type of projects you are working on and what your goals are. Take your time to find the perfect mic for you, and if it is on the more expensive side then get a basic mic and save up for the one you really want and deserve.

Your microphone will be connected via the lightning port. Using the adapter will allow you to connect your microphones USB connector to the adapter and adding the adapter to your phone, allowing you to now record on a different sound quality than just using the default built-in speaker that your iPhone or iPad comes with.

The adapter is called the Lighting to USB3 Camera Adapter, and while I can say that going for a cheaper alternative is an option, especially since Apple products are somewhat more expensive, it is wise to consider paying the slightly more expensive price. Apple is known for rolling out software updates that patch the charging ports to not support unofficial adapters, rendering your alternative adapter or charging cable moot. If at first you cannot afford the Apple official adapter, you can opt for an alternative that can be found on Amazon or Ebay, but be sure to constantly check the software updates and make sure to save up so you can get the adapter as soon as possible. What makes this adapter so amazing is that you can connect various devices to your iPhone and/or iPad, not just a microphone or audio interface.

To use your microphone, simply plug the microphone's USB plug into the adapter and then plug the lightning adapter into your iPad or phone. It is as easy as that. If GarageBand is open when you make the connection, a pop-up window will appear, letting you know that the software has detected a microphone, simply tap the 'Turn on monitoring' selection and start your recording journey. The pop-up will also inform you that wearing headphones is the safest option as it will avoid feedback when playing back the soundtrack.

Most of the newer Apple products no longer have headphone jacks, so this might become an issue, as there is no space for you to plug headphones in, and while you can connect bluetooth headphones to monitor your recording, you will get a pop-up that lets you know that there will be a delay in the playback and it might cause some problems later. The other alternative that you can opt for is to buy a microphone that has what is called an onboard headphone jack. This will allow you to connect your headphones directly to the microphone and not your iPad or iPhone.

Finding a great budget microphone is different and while I can suggest ten different ones I would opt for the Blue Snowball microphone. It unfortunately does not have an onboard headphone jack, but it is small and compact, meaning you can travel with it easier than normal. It also connects easily to the adapter and doesn't cause issues with connection or delay. If you want to spend a little more money you can opt for the Blue Yeti as it is a little bigger and allows more functionality, or the Samsung G-Track Pro. While both of these options are a little more expensive their capabilities and quality are out of this world.

The Samsung G-Track pro includes an onboard headphone jack for monitoring and has its own dials and controls on the mic for even more control over the quality and sound you create. While this is the best microphone to be using, it is not a plug-and-play mic at all. You will have to change some settings to be able to use the mic with your iPhone or iPad. This specific mic, like with the audio interfaces we covered earlier in the section, requires more power to be able to work. This means you will be using the second lightning plug in your adapter to plug your charger into, to give the microphone enough power to work. Once that has been done you can get back to recording and singing to your heart's content.

Quantization

When you start your recording process on iOS GarageBand you might find yourself struggling to keep up with timing and sometimes you might even fumble. This is perfectly normal and it happens to the best of us. While this does get better with time, GarageBand offers its users the capability to fix these small mistakes and timing issues simply by using quantization.

In this section I will cover how to use quantization on your tracks in the iOS Garageband software. As with most things in life, timing is extremely important no matter what, and to create a cohesive and successful track, you will have to make sure that your timing is on point. While adjusting and changing timing in MacOS GarageBand is somewhat easier, moving to a smaller screen where you have to keep track of timing and playing an instrument on a small screen and recording, it is much harder to get your timing perfect each and every time. That is perfectly fine though.

GarageBand has a wonderful thing called quantize. This is a simple tool that can be used to correct small timing mistakes between yoru various tracks. To quantize your track you will need to solo your other track/s first, allowing you to only change the settings on one specific track. You can solo your track by tapping the headphones icon next to the track that you will be working on. Next you will want to open the settings pane. You can do this by dragging the track icons to the left, this way pulling the menu open, or you can use the adjuster icon in the top-left corner of the LCD. Once the settings menu on the side has opened, simply tap on the 'track settings' menu and then on the quantization menu, you can now choose from four different options.

- 3. None
- 4. Straight
- 5. Triplet
- 6. Swing

Depending on the instrument and recording quality that you are working with, selecting the straight option with a 1/16th note selection will be perfectly fine for simple beats. The note selection in this section will create quantization for every 16th note, but if this is not to your liking and you think a tighter sound would be better for your track, you can opt for the ⅛ notes instead. This simply means that once every 8 notes the quantization will look for the timing and fix it if it does seem to be out of par.

As with all of the other settings and adjustments that can be made, I would suggest playing around with the different settings. While there are specific ways to do certain things the best way to find what works for you is by playing around with the settings and seeing how you feel about it.

Time Keeping

Time keeping in GarageBand is something that not everyone can do right off the bat and while GarageBand offers up various ways to do that, being able to do so yourself will allow you an easier and better time recording instruments and vocals.

The easiest way to get better at time keeping is to exercise your capabilities. This sounds far more menacing than it ought to be. In the next section I will highlight a few exercises that you can try and continue to do to get your timekeeping better, especially when it comes to using GarageBand.

One of the easiest ways to tell if you are keeping in time and tune is to record yourself. While this might seem daunting, you will far easier pick up your timing issues when listening to yourself. When listening to yourself talk, you listen more critically. This critical listening will allow you to pick up on subtle and small time issues throughout the song or recording. You will be able to tell when you have been rushing through certain sections or you might find that instead of correcting small note mistakes you would just play over the mistake and continue on. Recording yourself and listening to it will allow you to critically think about the sound you are producing and consider how to change it.

When you are rushing through some sections, perhaps try playing those sections slower at first, practice the slower version, doing it over and over and over until you feel comfortable enough to move to a faster pace. This will allow you to critically engage with the content you are creating and not just mindlessly create something you aren't really proud of.

A second way to learn how to keep better timing is to set up a metronome that you can actually see while you are playing. While GarageBand does offer this to an extent, the sizing is far too small to make an impact. I would suggest finding an app or online platform that uses a flashing or blinking metronome where you can switch off the ticking or clicking sound of the metronome you would use.

By switching off the sound and focusing on the flash or blink that is created you will be able to move away from a headspace where you are actively listening to the metronome. Some artists have mentioned that they often get stuck waiting for the metronome click to signal that they are playing at the right pace or time, but this leaves you unfocused on the task at hand. Simply forcing yourself to rely on the visual queue rather than the audible one will allow you to focus more on the sound you are producing than playing a waiting game with the metronome.

The final time keeping exercise that I will suggest is for you to try to better yourself is to work on your limb independence. It is perfectly normal that one of your hands or legs, depending on the instrument you play, is somewhat less practiced. This is a common occurrence, especially for instruments that use both your dominant and non-dominant hand. The best way to get this on a good level of time-keeping is to make sure that both limbs are equally dominant. There are various ways to do this.

Some artists will practice this by having one hand do one signature whilst the other hand does another signature. Using hand exercises are also extremely helpful as they will help your non-dominant hand become used to taking the same amount of strain as the other.

TIPS AND TRICKS FOR MAC GARAGEBAND

This section of Chapter 1 will cover basic tips and tricks to keep in mind when using Mac GarageBand. Some of the information will include content that was not covered in the first part or in this part. While some of this information will be new it will mainly include basic usage when it comes to certain effects and settings.

Automations

Automations are an easy way to create certain changes and effects in tracks. They also allow you to create automatic changes in tracks. Automations include things like creating automatic volume changes or fade outs. In the following section I will cover three basic automations that every beginner should use in their tracks.

Automatic Fade Outs are an amazing thing and having the software do it for you, without you having to worry about it when you have been spending hours crafting a new soundtrack is tedious and exhausting. The easiest way to get GarageBand to do this for you is to create an automation for it. Simply open the toolbar at the top of your screen and click on the Mix menu item. A small menu will appear, simply click on the 'Create Volume Fade Out on Main Output'. Once you have selected this GarageBand will automatically open the Master track as well as the automation view. Here you will be able to see the automation curve slowly pointing downwards, letting you know that the track will have an automatic fade out. You can adjust these settings simply by dragging the automation curves around in the track, this will either shorten or lengthen the fade out timing.

You can also create these curves at the start of a track, creating a fade in. However since there is no menu option for this you will have to create the curve yourself, either by adding in all the automation points yourself or by copying the fade out curve and duplicating it in the track and then manually moving the points to create the upwards curve.

This automation technically can also be used for track level volume fade in and fade out, allowing you to fade a track in that might only start halfway through the song, instead of having it blare out the moment it starts.

The second thing that automation is fantastic for is the pan knob, while this is slightly unconventional it is something I enjoy doing quite a bit as it moves the sound from one speaker to another while the track is playing. To do this you will need to open the automation panel simply by hitting the A key on your keyboard. Once this has been done, select the track you want to add the panning automation to and use the dropdown menu to select pan. Now you will be adding your automation points. Depending on where you want to start I would suggest adding a automation point at the bottom start of the track loop, and then once again in the middle of the track loop, but this time in the middle, a third automation point will be added on the last edge of the loop, and at the bottom to create the panning effect that the track is moving from one speaker to the other.

This is specifically used in films when there are view scenes that change or when the camera is slowly circling a large open area. It creates the feeling that you are hearing the sounds that are being made as the camera moves, giving a wonderful point of view feeling to the screen. This is also used by ASMR creators who want sound to slowly move from one ear to the other.

The final automation that can be done to create a more dynamic track is to change and automate effect parameters. These settings will automatically and take away effects on certain areas of your project and tracks. For the purposes of this exercise I will explain how to animate a growing filter effect on a track. The first thing you will need is to open the master track. You can either use the toolbar to open it, by clicking on the track menu and selecting 'Show Master

Track" or you can use the shortcut which is Shift, command and M used together.

Once the master track has been opened you can open the smart controls panel at the bottom of the screen and the plugins menu will be located on the left side of the project. For this example I will use the autofilter plugin that can be found in the GarageBand plugins under filter. Now it is time to open the automations window, simply by hitting the A key. Now select the master track to add the automation to and use the dropdown menu to select the output menu and then the Autofilter menu item that we used. The option will pop-up that we can automate different parts of the autofilter and for this example we will use the 'cutoff' selection.

So first add an automation point at the start of the track, somewhere in the middle to have the volume be at around 25 percent, then add a second automation point at the end of the first 8th bar, or when the music starts playing. Drag the second automation point to the very top of the track, setting the filter cutoff at 100 percent.

Master Tracks

The master track is the final track that can be edited and changed to your heart's content. It is the track that will be exported once you are done. Changes to the master track will be ultimate changes and I suggest only making changes to the master track that you want to keep in the long run, such as fade outs or automations.

As with any of the changes and edits, I suggest duplicating the track or project to make sure you do not ruin your track or lose data. In the following section I will share some interesting tips when it comes to the master track and how changes you make will make the quality of your final product be even more astounding than before.

The master track is like the final product where you can adjust settings after you have completed creating all the sounds and loops and automations in the other tracks that become the master track.

To view the master track simply use the shortcut or the toolbar tracks menu.

One of the easiest and best ways to utilize the master track is to use the Master track as a way to master all the other tracks. This will allow you to control the final output much more closely and with a more detailed eye. The best way to master your tracks with the master track is to export your tracks or mix to a WAV or AIFF file and then use that singular track and import it into a new project. This allows you more creative freedom when adding things such as EQ, compression and many other effects.

This not only helps you have better control over your overall track output but also gives your computer a break, letting off some of the CPU pressure from having to constantly work on multiple tracks might help your computer make changes faster or do so without freezing or even crashing.

While you do not have to export your file, and you can work on your Master track alongside your other tracks for convenience sake, I would still suggest working on it separately, especially if you have an older model Macbook or iMac that might struggle with the newer software and end up crashing on you. There is nothing worse than working for hours on end on a project and having the software crash on you only for you to have to redo that work. Make sure you are saving constantly, so that if for some reason you lose power or the software crashes you do not lose power.

When you have your master track loaded, you can open your library and add any one of the 9 different factory plugins to your track. These plugins are defined using the following terms:

- Ballad
- Classical
- Dance
- Hip Hop
- Jazz

- Pop
- Rock
- Top 40
- Experimental

Each of these factory plugins has its own settings and effects that get added to the master track, giving the track a very great overhaul feeling.

The second thing that master tracks can be used for and that are a great addition to any track is a fade out. As I covered the process in the automations section above I will only briefly discuss this. A fade out, whether at the end of a track to create a wind down or when fading out a certain instrument is always a wonderful addition to the track.

Using your master track to add fade outs will allow you far more control over how the sound will be exported, instead of having to create fade outs for each track and lining them up. Fade outs also go hand in hand with fade ins, allowing you to let a certain sound slowly fitler into the entire track. These however are much better suited being attached to the specific track as adding a fade in to the master might create a random drop in volume in the middle of the song and nobody wants that.

The final thing that I love about the master track and the manipulations one can achieve with it is by adding some dynamic effect plugins to the master track. What makes this even more amazing is the fact that GarageBand offers a wide variety of effect plugins by default that can be added to the master effect to create some amazing sounds.

Each effect will add different settings and changes to your master track, letting the feeling and ambiance change. Space Designer is one of my favorite effects to add to any tracks as it creates an eerie space vibe. As with any of the track edits and changes I would suggest playing around and spending some time doing research about more effect plugins that can be downloaded and installed into GarageBand to offer you an even bigger variety of track creations.

Exporting Your Projects

Now that you are finished editing and changing and shuffling tracks around and adding effects and you are happy with the product that you have created you can finally move on to the export section of the project. It is exciting considering that you just created an entire song, or an audiobook, or even recorded narration for an entire film. You can be proud of yourself for working so hard and pushing through the long nights and the struggles of finding the right settings for you and your project.

There are various ways to export your projects and various platforms where you can add and upload your projects to be able to share them with your family, friends and the internet. Platforms like SoundCloud, Apple Music, the Play Store and even Youtube are all places where you will be able to upload your projects.

GarageBand offers you a wide variety of sharing options, listing 7 different ways to share your projects with anyone. There are also some iCloud specific sharing selections that can be made, but since the focus is on GarageBand rather than MacOS, I will be excluding information on these methods.

While most of these are pretty easy to understand and comprehend, there are some steps that might need some clarification. These sharings options are found underneath the share menu in the toolbar and are named as follows:

Song to iTunes

When sharing a song through this method the song will show up in your Macbook or iMac's Apple music library. When hitting the 'Song to iTunes' button you will be prompted by the software to fill in a form with information such as the title, the artist name, the composer name, the album name, adding it to an iTunes playlist as well as selecting the quality you want to use when exporting the track.

The quality selection also goes hand in hand with the compression into MP3 format. Uncompressed audio files take up more space as they include all the track data.

Song to Media Browser

When you select this option your project will directly be exported into GarageBand's own media browser. This allows you to access the track in any other browser that is related to other GarageBand projects. This option is very rarely used and not a lot of people stand by it.

People generally opt to simply drag and drop files directly from Finder into GarageBand instead of using the media browser.

Song to SoundCloud

This option allows you the easiest way of sharing your track to SoundCloud. When selecting this option you will get a similar dialogue box as with the iTunes selection.

However here you will need to select a source file, set the visibility and permissions for the song and if you are not logged into your SoundCloud account you will be prompted to do so before moving onto the dialogue box. Once this has been done you can simply click the share button and the rest will be done for you.

AirDrop

AirDrop is Apple's version of sharing something via Bluetooth, however you need to be sure that your Wifi and Bluetooth is switched on for both the device you are sharing from and the device you will be sharing to.

When selecting to share your project like this you will be prompted to choose between either sharing the project, which will allow for collaboration, or the song.

When sharing to your iPhone you will not be able to open this file if you choose to open it in iOS GarageBand, to be able to do that you will have to use option 7 first before AidDropping the file to your iPhone.

Mail

When sharing the project via Mail the same dialogue box prompt will pop-up as when you are sharing via AirDrop. You will need to choose between sharing the project file, or sharing the song, before choosing the quality of the file and then sharing it. GarageBand will then open your Mail and automatically attach the file or song to an email body and you can fill in the rest. Again, you will not be able to open the file in iOS GarageBand as the file extension is incorrect.

Export Song to Disk

Sharing the project or song to the Hard Disk will save the file somewhere on your computer. When exporting in this manner you will be able to manually add it to any of the music streaming apps such as iTunes, SoundCloud, Spotify and Youtube. The Dialogue box that will pop up will require you to select a file type as well as giving you the accessibility of being able to choose where exactly you want the file to be saved. One of the most common file types is MP3, however WAVE is also a dynamic file extension. Choosing to export the file in MP3 allows you to select the compression quality of your track so I generally opt for that. It is also a widely recognized format and most streaming and music applications will allow this file type.

Project to GarageBand for iOS

I covered this sharing method earlier at the end of Chapter 4, but will briefly discuss it again. Sharing your file in this way allows you to AirDrop or Mail the mobile.blend file to your iPad or iPhone. This is the only format that is accepted by GarageBand in terms of unfinished GarageBand projects. If you choose to export your file in AAC format to your Hard Disk you will be able to import it into GarageBand but you will not be able to make any changes to the file in any way whatsoever. When using the appropriate file you will be able to add edit and change the track as much as you want and then simply save and share it back to your computer to finalise in Mac GarageBand.

It is important to note that when sharing the project back to make the final adjustments on Mac GarageBand that you select share as Project, as any other format will compress the file and you will not be able to make any other edits.

Final Thoughts

While all of these tips and tricks are here to help you better understand the mobile version of GarageBand, I truly hope and believe that you can take these skills and translate them into using MacOS GarageBand. While you will no longer be tapping your screen when you use MacOS GarageBand, the knowledge and skills that you have learned for iOS GarageBand will stand and allow you to create better tracks in the future.

I have said this multiple times in both this part and in the first part but the easiest way to get to terms and grips with how both Mac and iOS use GarageBand is to practice and practice and practice. This will be the only way to learn how to use the software efficiently. When I started my journey with both these platforms I barely knew the basics and I played around a lot within the software, recording random and strange sounds around my house and area and adding different effects and settings to those tracks helped me understand how the settings and effects worked and how they would impact certain sounds and how other sounds would not be changed at all.

Don't be afraid to try new things and to explore sound creation, there is no wrong or right way to create. "Imagination is more important than knowledge. Knowledge is limited. Imagination encircles the world."(Einstein, 1931)

I hope that your journey up until now has been a pleasant one. I believe that all your hard work up until now will be realized and the new skills that you have learned will allow you to follow and chase your dreams to your heart's content.

Feel free to discover even more surrounding music production, music lessons as well as singing with our wide range of guides and tutorial booklets.

CONCLUSION

Whether or not you're just learning GarageBand or it is a product you have worked with before you can now rest easy knowing that the information provided in this book will make it much easier for you to use both the MacOS and iOS versions of GarageBand.

In Part 1, the first chapter 1 covered the basics for those who have never used the software before and needed a small introduction to GarageBand and what it can offer them. Chapter 2 covered the basics every producer needs to know about creating a new project from scratch, including the type of tracks and sounds that you can add to your projects. Chapters 3 and 4 covered how to use those skills to create basic tracks and how to include a film into your project. Chapter 5 and 6 covered the best ways to enhance and refine your project to create the perfect blends for your signature sound. Chapter 7 covered the final checklist items you may want to check out before finishing off your project, and Chapter 8 covered Apples' amazing Lesson section where you can learn how to play either guitar or piano.

The mini-tutorials were a way for you to familiarize yourself with the material in each chapter and I hope that you now feel comfortable enough to play around inside the software on your own. The best way to learn a new trade is to practice, practice, practice.

In Part 2, Chapter 1 was a quick recap of all the information that was included in the first part of this series, quickly going over all the important information you needed to know about the basics, as well as shortcuts and usage.

Chapter 2 covered all the information you needed to know about Projects and recording as well as all the information that you need to focus on when adding new loops to your project. The tutorial in Chapter 2 covered how I go about starting any new project and what steps I follow to get through the basics and the boring settings.

Chapter 3 covered all the content regarding Arrangements and how their settings and adjustments will help you be a better music and audio producer. The tutorial included my step-by-step guide on how to make adjustments to the tracks that you have included as well as how I go about my selection and change process as well as my editing process.

Chapter 4 covered all the information with regards to smart controls, amps and pedals. In this chapter I covered the edits I like and I included some recommendations with regards to the plugins I use. These plugins are amazing and give any track the much needed depth they desire. The tutorial I gave in Chapter 4 covered how I go about my editing process, and I included information on exporting your project to use in the iOS GarageBand or exporting it for publishing purposes.

Chapter 5 covered GarageBand on iOS and all the basics surrounding it. I tried to include as much information as possible to allow you an easy way to keep editing and creating your tracks without having to carry your computer around with you all day. The tutorial in this chapter covered how I would import my track from my computer to my phone and how I would continue editing while on the go.

The tutorials in this book were designed and created to give you the basic skill set you might need to be able to grow and educate yourself surrounding the process while also giving you the creative freedom you deserve. I hope that the tutorials will help you in your endeavors to become a great production artist. I have learned that the best way to master something is to do it over and over again until it is perfect and then doing it again.

Now that you have all this knowledge and you are ready to jump into the world of custom track creation I hope that you can use all the knowledge from this book to create amazing music and audio!

I hope that this journey has been as amazing for you as it has for me! The information covered is not the only information you will find and I highly suggest that you keep playing around and practicing as much as you can and want to. It is important to keep trying new things, as there's a plethora of unknown when it comes to every type of setting and capability. Good luck and have fun!

SHORTCUTS

All of the following shortcuts can be used within GarageBand. They will be split into three categories. Main Window shortcuts, Editor Shortcuts, and Global Track Shortcuts.

Main Window Shortcut Keys

These shortcuts are used in the main window as referenced above but can also be used in the Editors pane.

Shortcut	Action
R	Starts the recording
Space bar	Starts or stops the playback
K	Turns the metronome on or off
Command + Q	Quits GarageBand
Command + N	Opens a new project
Command + O	Opens an already existing project
Command + S	Saves the current project
Shift + Command + S	Saves the current project as ...
Command + Z	Undo
Shift + Command + Z	Redo
Command + X	Cut

GarageBand Basics

Shortcut	Action
Command + C	Copy
Command + V	Paste
Command + A	Selects all
Enter (numeric keypad)	Starts the playback of your selected track
0 (Zero - numeric keypad)	Stops the playback of your selected track
, (Comma)	Moves the playhead back in your track by one bar
. (Full-stop)	Moves the playhead forward in your track by one bar
Shift + Space Bar	Starts the playback from your currently selected track
Shift + Command + .(Full-stop)	Moves the cycle range onward by one full cycle length
Shift + Command + ,(Comma)	Moves the cycle range backward by one full cycle length
Function + Left Arrow Key	Goes to the beginning of the current selection

Shortcut	Action
Option + Backspace/Return	Goes to the end of the final region
Return/Backspace	Goes to the beginning of your sound
C	Toggles any cycle area on or off
Control + Option + Command + S	Toggles solo on and off for any solo tracks
K	Toggles the metronome on or off
Shift + K	Toggles the count-in on or off
Command + ,(Comma)	Opens the preferences window for GarageBand
Command + K	Shows or hide the musical typing window in
B	Shows or hides Smart Control windows
N	Shows or hides Score Editor
P	Shows or Hides Piano Editor

Shortcut	Action
O	Shows or Hides Browser for Loops
Y	Show or Hides the Library section
Shift + / (Backslash)	Show or Hides the Quick Help section
Command + / (Backslash)	Shows detailed Help section
Control + Command + F	Toggles Full-Screen mode on or off
Option + Command + O	Opens the Movie window
Command + W	Closes current window or project
Up Arrow Key	Selects next higher track
Down Arrow Key	Selects next lower track
Command + H	Hides GarageBand
Option + Command + H	Hides any other open applications

Shortcut	Action
Tab	Changes the focus to the next area in the GarageBand window
Shift + Tab	Changes the focus to the previous area in the GarageBand window
Command + M	Minimizes the GarageBand window
Option + Command + M	Minimizes all other open application windows
]	Moves to the next Patch, Effect, or Instrument setting
[	Moves to the previous Patch, Effect, or Instrument setting
M	Toggles Mute on or off for the selected track
S	Toggles Solo on or off for the selected track
Control + I	Toggles monitoring of a track on or off
Delete	Delete
Command + Left Arrow Key	Horizontally zooms out
Command + Right Arrow Key	Horizontally zooms in
Function + Up Arrow Key	Pages Up

GarageBand Basics

Shortcut	Action
Function + Down Arrow Key	Pages Down
Function + Left Arrow Key	Pages Left
Function + Right Arrow Key	Pages Right
A	Shows or Hides the Automation lanes
Option + Command + N	Creates a new track
Option + Command + A	Creates a new audio track
Option + Command + S	Creates a new software instrument track
Option + Command + U	Creates a new Drummer track
Command + Delete	Deletes the selected track
Left Arrow Key	Selects the previous region on the selected track in the Editor window
Right Arrow Key	Selects the next region on the selected track in the Editor window
L	Continuously Loops the selected region
Q	Quantize the selected events
Option + Command + Q	Undo Quantization of events

Shortcut	Action
Shift + Command + V	Pastes and replaces the current selection
Command + J	Joins any regions or notes
Command + T	Split the selected region or event at the playhead position
Command + G	Toggles the 'Snap to Grid' on or off
Control + Shift + Command + Delete	Deletes all the automation on the selected track/s
Control + R	Toggles Record Enable for the selected tracks on and off
Option + Command + Delete	Cuts the selected Arrangement Marker into various sections
Shift + Return	Renames the selected track
Control + Shift + O	Adds the selected region to the Apple Loops Library
Shift + N	Renames the selected regions
Option + T	Configures the Track Header for the selected track
E	Shows or hides the Editor pane
Option + Command + P	Shows or hides the notepad pane

Shortcut	Action
F	Shows or hides the Browser pane
Shift + Command + M	Show or hides the master track for the project
Option + Command + G	Show or hides the alignment guides for the project

Editor Shortcut Keys

The editor shortcut keys are used when you are using any of the editors that GarageBand offers.

Shortcut	Action
Options + Space	Preview the currently selected audio
Options + Up Arrow Key	Transposes the selected notes up by one semitone
Options + Down Arrow Key	Transposes the selected notes down by one semitone
Options + Shift + Up Arrow Key	Transposes the selected notes up by one octave
Options + Shift + Down Arrow Key	Transposes the selected notes down by one octave

Global Track Shortcut Keys

These shortcuts are used to either hide or show any of the arrangement, video, or transposition tracks as well as the tempo tracks.

Shortcut	Action
Shift + Command + A	Shows or hides the arrangement tracks
Shift + Command + O	Shows or hides the movie or video tracks
Shift + Command + X	Shows or hides the transposition tracks
Shift + Command + T	Shows or hides the tempo tracks

THANKS FOR READING

Dear reader,

Thank you for reading *GarageBand Basics*.

If you enjoyed this book, please leave a review where you bought it. It helps more than most people think.

Don't forget your FREE book chapters!

You will also be among the first to know of FREE review copies, discount offers, bonus content, and more.

Go to:

https://offers.SFNonfictionBooks.com/Free-Chapters

Thanks again for your support.

REFERENCES

Apple Support. (n.d.). *MacOS App User Guides: GarageBand for MacUser Guide*. Apple Support. Retrieved July 31, 2021, from https://support.apple.com/en-za/guide/garageband/welcome/mac

Audio Keychain. (n.d.). *Song key finder | Your source for creating the perfect mashup*. Audio Keychain. https://www.audiokeychain.com/

Baird, P. (n.d.). *Getting started with GarageBand for iOS and iPadOS landing page true*. The GarageBand Guide. https://thegaragebandguide.com/getting-started-with-garageband-for-ios-and-ipados-landing-page-true

Clark, B. (2020, June 29). *10 great free GarageBand plugins & how to install them*. Musician Wave. https://www.musicianwave.com/free-garageband-plugins/

Duncan, L. (2018, March 1). *6 best exercises to improve your timekeeping on any instrument*. Music Industry How To. https://www.musicindustryhowto.com/6-best-exercises-to-improve-your-timekeeping-on-any-instrument/

Einstein, A. (1931). *Cosmic Religion and Other Opinions and Aphorisms*. Literary Licensing.

Reed, G. (n.d.). *You are being redirected...* Ethos3.com. https://ethos3.com/2017/03/microphone-test-phrases-every-presenter-should-know/

The GarageBand Guide. (2020a, March 25). *5 Hidden GarageBand tricks* [Video]. Youtube. https://www.youtube.com/watch?v=yFWZeCXEPsQ

The GarageBand Guide. (2020b, July 23). *GarageBand iOS for beginners - YouTube* [Video]. Youtube. https://www.youtube.com/playlist?list=PLa_TA4RKaD9OWMJphKb7-CpFTpVjHM7Iq

The GarageBand Guide. (2021, February 22). *Best free GarageBand plugins iOS* [Video]. Youtube. https://www.youtube.com/watch?v=AW9-01pvVQE

TheGarageBandGuide. (2019). *GarageBand tutorial: How to export your projects* [Video]. YouTube. https://www.youtube.com/watch?v=_2UIZ90wXmw

All Images sourced from GarageBand, via Screenshots, and resketched by Neil Germio.

AUTHOR RECOMMENDATIONS

Play in Perfect Harmony

Discover how to express yourself through rhythm and notes, because music theory doesn't have to be intimidating or tedious.

Get it now.

www.SFNonfictionBooks.com/Music-Theory-Beginners

Discover How to Use Yoga as Medicine

Discover how to heal yourself naturally with *Curing Yoga*, because you deserve to feel your best.

Get it now.

www.SFNonfictionBooks.com/Curing-Yoga

ABOUT AVENTURAS

Aventuras has three passions: travel, writing, and self-improvement. She is also blessed (or cursed) with an insatiable thirst for general knowledge.

Combining these things, Miss Viaje spends her time exploring the world and learning. She takes what she discovers and shares it through her books.

www.SFNonfictionBooks.com

amazon.com/author/aventuras
goodreads.com/AventurasDeViaje
facebook.com/AuthorAventuras
instagram.com/AuthorAventuras